A SHEARWATER BOOK

For the Health of the Land

ALDO LEOPOLD

For the Health
of the Land

*Previously Unpublished
Essays and Other Writings*

*Edited by
J. Baird Callicott and Eric T. Freyfogle*

*Foreword by Scott Russell Sanders
Afterword by Stanley A. Temple
Illustrations by Abigail Rorer*

ISLAND PRESS / Shearwater Books
Washington, D.C. • *Covelo, California*

A Shearwater Book
published by Island Press

Copyright © 1999 Island Press

Illustrations copyright © 1999 Abigail Rorer

Shearwater Books is a trademark of
The Center for Resource Economics.

Library of Congress Cataloging-in-Publication Data

Leopold, Aldo, 1886–1948.
For the health of the land : previously unpublished essays and
other writings / Aldo Leopold ; edited by J. Baird Callicott and
Eric T. Freyfogle ; afterword by Stanley A. Temple.
p.   cm.
Includes bibliographical references.
ISBN 1–55963–763–3 (cloth)
1. Nature conservation. 2. Landscape protection. 3. Natural
history—Outdoor books. I. Callicott, J. Baird.
II. Freyfogle, Eric T. III. Title.
QH75.L38   1999
333.7′2—dc21   99–16797
CIP

Printed on recycled, acid-free paper

Manufactured in the United States of America

10   9   8   7   6   5   4   3   2   1

For Charles Bradley,

a healer of land

Contents

Part III: Conservation and Land Health

Foreword: Reading Leopold

Scott Russell Sanders

W<small>HY DO SO</small> many of us seek out the company of Aldo Leopold? Nowadays we can commune with him only by reading his books, of course, or by visiting those sandy acres in Wisconsin that he helped reclaim from ruin, or by remembering him as we hike the tracts of wilderness that he helped preserve. But whether he is encountered on the page or in memory, he remains an abiding presence. No matter where I go, when I fall into conversation with people concerned about how we ought to live on this imperiled and magnificent planet, Leopold's name comes up. He is one of the touchstones in our thinking about nature and culture, one of the essential figures we reckon by. What keeps him so vividly present in our minds more than fifty years after his death? I cannot speak for others, but I can try to say why I rejoiced when I learned about this new collection of Leopold's essays.

The essays reached me by mail in early April as I was leaving my home in Indiana for a journey to Virginia, so I carried them with me to read on the way, trusting that Leopold's voice would be an earthy counterpoint to the drone of airport loudspeakers and jet engines and hotel air conditioners. And so it proved to be. In these pages I found the same vigorous, humorous, practical, exuberant companion whom I had long relished in *A Sand County Almanac.* "The best anti-gopher insurance," he tells us, "is a pair of nesting redtail hawks in the woodlot." "Label each notable tree," he advises us, "not with pieces of tin or wood nailed to the bark, but with pieces of thought and understanding nailed to your mind." "The damage done by birds," he reminds us, "is like that done by dogs or children; if you like them well enough there are ways to get along."

"When the farm boy can no longer see a fox-track in the snow," he says, "it will take a lot to replace the loss." Here is a guide who concedes that foxes make occasional raids on chicken coops but who then goes on to defend all predators: "The ultimate question is whether a foxless, hawkless, owlless countryside is a good countryside to live in. Some of us, for reasons that wholly defy explanation, are sure it is not. Some of us have seen countrysides from which all these really wild things are gone, and in which land-minded people regret their going, and make pathetic efforts to restore them." The man we meet in these pages is thoroughly land minded; he is an unabashed lover of wild things. If we share even a hint of that love, we are bound to enjoy his company.

I began reading these outdoor essays in a crowded airport lobby where television monitors poured out advertisements thinly disguised as news and travelers shouted into cell phones about schedules and dollars. While I was trapped inside this noisy box, gazing out through plate glass windows on a vista of concrete and huge machines, what a relief it was to hear Leopold celebrating the fox, the frog, the prairie grasses, the wild tangles of grapevines, the musical flight of geese, the rising sun, the coasting moon, the quicksilver clouds. How reassuring it was to hear Leopold's wise voice countering the "slick and clean" ideal of industrial farming with his own ideal of the farmer as "a husbandman of living things." A day spent traveling by plane should convince you, if you do not already know, that we live in an increasingly slick and clean world where nothing is husbanded except money and power, nothing valued except efficiency and speed.

As I waited in line to board the plane for my flight to Virginia, a 747 came wheeling in to the next gate, and no sooner had it stopped than a flock of sparrows swirled down to hunt for tidbits among the massive wheels. As luggage carts and fuel trucks roared up, the sparrows flittered aside for a moment and then settled again to resume their search, hopping about, spry and quick. They were like tiny flames that even a giant's breath could not put out.

Watching them, I thought of Leopold's praise for the bog-birch, "such a mousy, unobtrusive, inconspicuous, uninteresting little bush," and his praise for the bur oak, "the only tree that can stand up to a prairie fire and live." But I also recalled his loving descriptions of creatures less adaptable than the bur oak and the plucky sparrows: prairie chickens and long-eared owls, sunflowers and ferns, river otters, wolves, and countless other species that were perishing beneath the wheels of our relentless enterprise.

From the airport, I rode toward Charlottesville past redbud trees glowing the color of burgundy wine, the white bursts of shad-bush, and the creamy candelabra of dogwood, all blooming with the urgency of April. But nearing the city I also passed one building site after another where bulldozers tore up the woods and earthmovers scoured the hills, and signs announced the coming of discount stores, drive-in pharmacies, video arcades, gas stations, motels, curio shops. You cannot travel far in our prosperous country, nor read far in Leopold, without being reminded of loss. "Few acres in North America have escaped impoverishment through human use," he tells us soberly. "If someone were to map the continent for gains and losses in soil fertility, waterflow, flora, and fauna, it would be difficult to find spots where less than three of these four basic resources have retrograded; easy to find spots where all four are poorer than when we took them over from the Indians."

Drawing up his own maps during and immediately after the dust bowl years, Leopold did not have to look far for evidence of loss: "The symptoms of disorganization, or land sickness, are well known. They include abnormal erosion, abnormal intensity of floods, decline of yields in crops and forests, decline of carrying capacity in pastures and ranges, outbreak of some species as pests and the disappearance of others without visible cause, a general tendency toward the shortening of species lists and of food chains, and a world-wide dominance of plant and animal weeds." Leopold was among the first observers to tell us in scientific detail that our

seemingly robust land is ailing, and no one has told us more con-
vincingly. Of course, even today many people—including those
who order the bulldozers and earthmovers into motion, those who
market gasoline and cars, those who make money from handling
money—are still not convinced that anything is amiss. They refuse
to believe that our headlong pursuit of wealth and pleasure could
diminish the earth's ability to support life. Such people will tell you
that any talk of land sickness is an attempt by doomsaying envi-
ronmentalists to halt progress.

As a visitor to Charlottesville, I found it hard to see how the citizens of that beautiful old city could bear to watch it bulge outward
along every highway, choke with traffic, and fill with fumes while
redbud and shadbush and dogwood fall before the bulldozers,
while forests and farms give way to mile after mile of pavement.
Leopold questioned the sort of "progress" that enriches the human
world by impoverishing the nonhuman world. He challenged the
belief in technological fixes, warning us about DDT, for example,
more than twenty years before Rachel Carson wrote *Silent Spring*.
He disputed the views that bigger is better, that novelty is proof of
vitality, that profit matters above all else. His skepticism about the
dogma of endless growth is another quality that makes him seem
like our contemporary. I suspect that he would have been appalled
by the bloating of Charlottesville and by the aimless swelling of
nearly all American cities, including my own hometown of Bloom-
ington, Indiana.

Before I return home from this little excursion in praise of Aldo
Leopold, let me mention the last stop on my trip to Virginia, which
was at Monticello. As I walked through the grounds of Thomas
Jefferson's estate, among the terraced vegetable gardens and flower
beds, among the great trees, both native and exotic, I thought of
how many traits Leopold shared with the designer of this orderly
place. For all their learning, they were both practical men. Jefferson kept a workbench in the room adjoining his study so that he
could try out his designs by making models. Leopold's workbench

was the land itself, especially the arboretum at the University of Wisconsin and the sandy acres surrounding his shack beside the Wisconsin River. Both men loved to experiment; both loved to nurture growing things. Jefferson collected specimens of animals from the entire continent, many of them shipped back to Virginia by his two emissaries, Meriwether Lewis and William Clark, and most of them stuffed. Except for the occasional woodcock or deer or quail that he shot and gladly ate, Leopold preferred his animals alive and roaming free, but he shared with the master of Monticello an omnivorous curiosity about every wild creature. Neither man had much use for institutional religion, yet each one addressed the earth with reverence. Both looked to the small landowner as the best advocate of democracy and the surest defender of the land itself.

I draw this comparison not merely because I happened to visit Monticello while reading these essays but because I believe that Leopold and Jefferson form two points on a hopeful line running through American history. We Americans are all too aware of the cruelty and waste that have characterized much of our history, as hunters and pioneers and miners and merchants fanned out from the eastern seaboard looking for cheap land and quick wealth. Yet from the beginnings of European settlement there has also been a countermovement created by people less intent on making money than on making sense of where they were, curious and skillful people who chose to settle in one place for a spell or a lifetime and who studied their neighborhoods with a loving eye. There are far more figures in that tradition than I can name here, but they would include Michel-Guillaume-Jean de Crèvecoeur, John James Audubon, Henry David Thoreau, John Muir, and Rachel Carson from earlier generations, and Wallace Stegner, Gary Snyder, and Wendell Berry from our own time. Like Jefferson, Leopold is a key member of that promising lineage, which prepared the way for the ecological restorationists who are now working to heal injured lands from coast to coast, for the members of land trusts who are

protecting open space, for the teachers and children who are turn-
ing schoolyards into gardens, for the citizens who are cleaning up
rivers, for the legions of people who are striving to live with a simplicity that honors and welcomes wildness.

Were Leopold only a prophet decrying past sins, he would be
less vital to us, for we need more than a diagnosis of ills; we also
need a vision of remedies. He offers us such a vision throughout
his writing but nowhere more clearly than in the neighborly essays
collected in *For the Health of the Land.* I call them neighborly
because, in reading them, I imagine Leopold giving me advice
while the two of us walk over a patch of woods and meadows in
southern Indiana. It is land that my family has recently bought,
and with Leopold's help I can see the damage done by previous
owners: the eroded gullies, the stumps from careless logging, the
straightened creek, the rusted carcasses of refrigerators and trucks
dumped along the road. The gullies can be mended, he assures me.
The trash can be carted away. The creek can be slowed down and
encouraged to wander. Here is a field we could replant to prairie.
Here is a low spot that could be flooded for a marsh. Here is a snarl
of grapevines that could be trained so as to make ideal cover for
wildlife, and all the better if we plant some grain nearby and let it
stand through the winter. Here is a sunny corner we could set aside
for rare flowers. Here is a site for a bluebird house. With each
glimpse of a wilder future for this land, I enter more deeply into
the life of the place.

Although he proved to be excellent company during my journey
through airports, across the Blue Ridge Mountains, and down
among the dogwoods of Charlottesville and the towering tulip
poplars of Monticello, Leopold speaks to me most powerfully now
that I am back on my own home ground. On our imaginary walk,
he is wearing a soft old fedora, well-oiled boots, khaki trousers and
shirt; he carries a staff for pointing and poking as much as for
walking; binoculars and a dog whistle dangle from his neck; his
round spectacles glint as he turns to gaze with delight at every

mushroom and salamander and tree. He knows all the trees in our Indiana woods, all the flowers, all the birds, because he is a midwesterner, and that is yet another reason I am drawn to him. As much as he loved wilderness, Leopold also loved farm country, the mixture of cultivated fields, hedgerows, woodlots, pastures, homesteads, and meandering creeks that he knew from his childhood along the Iowa and Illinois margins of the Mississippi River, and that he knew as an adult in Wisconsin. Like Leopold's prose, and like the character of the man revealed in his books, the Midwest is plain and straightforward, yet shadowed with mystery; open and welcoming, yet reserved; full of promise and hope, yet acquainted with grief. The Midwest is a fertile, well-watered landscape, long inhabited, much abused and much adored, deeply scarred but also quick to heal if given half a chance.

Leopold's shorthand word for the effort of healing was *conservation,* by which he meant "a positive exercise of skill and insight, not merely a negative exercise of abstinence or caution." He insisted that conservation means more than simply preserving what has not yet been spoiled; it means reversing the history of abuse, using ecological principles to harness nature's own powers of recovery. He recognized that we have to make a living from the land, that we all need shelter and clothes and food. But he also realized that we need a great deal more if we are to lead sane and honorable lives: we need beauty, community, and purpose; we need "spiritual relationships to things of the land." Leopold understood what a radical claim this was in a society that sees land purely as real estate. He knew that neither fear nor scolding would move us to make such a profound change in our views of land and our ways of living, nor would logic, nor would law. We would be moved only by affection, by wonder, by joy in the presence of wildness—the very qualities that radiate from every page of this book.

For the Health of the Land

Introduction

J. Baird Callicott and Eric T. Freyfogle

THE ESSAYS in this volume were written by the person now widely acclaimed to be America's most perceptive and influential conservationist. They all deal with a crucial concern—how to promote conservation on privately owned lands. As the author saw it, that was the central conservation challenge of his generation. Today, our generation faces a wider suite of environmental problems, some of them global in scale. Nevertheless, more than a half century after Leopold's death, conservation of private lands remains a fundamental national concern. Nearly two-thirds of the land in the United States is owned privately.

The author of these essays, Aldo Leopold, is best known today for his book *A Sand County Almanac, and Sketches Here and There*, published in 1949, the year after his death. Translated into many languages, it has inspired millions of readers and is regularly cited as the most influential conservation book of the century, perhaps of all time. Leopold's writings, however, extend well beyond his *Almanac*. Many appeared originally in journals and newspapers that today are out of print and thus are accessible only to scholars. Others remained unpublished at his death, among them some of his most perceptive comments about the land community and the human role in it. Twelve items appear here for the first time ever; five of these are among the longer essays in parts I and III. Some of the essays draw on the particular circumstances of agrarian America at midcentury, but a wisdom and a passion transcending the time and place of Leopold's life run through them all. "How do we assess Leopold's words?" asks his biographer, Curt Meine.

In the half century since he wrote, conservation has evolved into environmentalism, while farming has moved toward agribusiness. Yet one need not read far into Leopold to appreciate the timeliness—or, perhaps more accurately, the time*less*ness of his thoughts. They remain relevant so long as people live on land and so long as the human instinct for stewardship endures.

The overall aim of conservation was a matter that Leopold pondered all his life. As a teenager, he wrote a school essay on timber famine and the need for forestry. In the 1920s, he focused on soil erosion and wilderness preservation. In the 1930s, his attention shifted to the restoration and management of "farm game" populations. But in the last decade of his life, he came to understand that sustainable forestry, erosion-control measures, wildlands preservation, and game management are not ends in themselves. They are parts of a larger, broader, more vital end: the maintenance of a healthy land community, a community that fuses human and other forms of life into a beautiful, fertile, and endlessly fascinating yet ultimately unfathomable whole. When late in his life Leopold spoke of the land and its conservation, he had in mind that entire bountiful, riotous, biotic community. "Who is the land?" he asked rhetorically in 1942. "We are, but no less the meanest flower that blows."

Although Leopold lived during the New Deal era and had many friends who worked for newly formed conservation bureaus, he remained convinced that government alone could never bring about effective conservation, particularly in landscapes dominated by private lands. Governments could pass laws and offer incentives, but successful conservation requires more than just regulation- and market-driven changes in outward behavior. Ultimately, it depends on landowners themselves. In the end, conservation requires shifts in the knowledge, skills, desires, and aesthetic sensibilities of the landowning populace and, ultimately, the development of a land ethic. Responsible landownership remains as important today as it was in Leopold's time. For landowners willing to take the lead, no guide is as true in philosophy or as good on the ground as Aldo Leopold, represented by the essays in this book.

One way that Leopold himself sought to promote conservation of private lands was by publishing a manual or book to help guide the land-use practices of farmers and other rural landowners. His aim was not just to get sound advice into the hands of landowners but also to instill in them a greater, broader love of all things natural, wild, and free and to help them see how a diverse, healthy, beautiful countryside could make their lives more enjoyable and their livelihoods more sustainable.

Leopold never completed that book, but he left behind the raw materials and a clear vision for it. Guided by that vision, we have assembled the pieces that seem most likely to fulfill his goals. These essays are vintage stock—clear, sensible, and provocative, sometimes humorous, often lyrical, and always inspiring. Leopold was writing not for other academics or scientists but for ordinary people, particularly farmers, and he wrote very well. He told his readers what they needed to know and why they needed to know it. He drew on his own long experience as forester, hunter, wildlife manager, pioneering ecologist, farm owner, and avid student of history and human nature. Here, one finds not only how-to advice on promoting game-bird populations, managing small woodlots, and fostering roadside prairies but also penetrating comments on nature—how its many parts fit together and how people are ultimately dependent on it and must work actively to conserve it.

From Wilderness to Wildness

Leopold wrote the essays in this book while living in the farm country of southern Wisconsin. He moved there in 1924 from the Southwest, where he had spent the first fifteen years of his career. He was thirty-seven years old, had a wife and four children, and worked for the United States Forest Service. The move brought him near his birthplace along the Mississippi River, in Burlington, Iowa, and it would be his family's last.

Why Leopold returned to his native bioregion remains unclear. By all accounts, he had come to love the rugged Southwest, still a U.S. territory and not yet the states of Arizona and New Mexico when he first arrived, in 1909. Certainly he had come to love, and had married, a young woman whose aristocratic family, the Lunas, had sunk roots in the soils of the Southwest ever since it had been part of New Spain. After a shaky start, Leopold had soon become a seasoned hand in the national forests of District 3—forests with such storied names as the Carson, the Apache, and the Tonto. In this magnificent, arid, sparsely settled country, he hunted and fished when he was not cruising timber and inspecting range.

During a wartime lull in forestry work, Leopold left the Forest Service to direct briefly Albuquerque's chamber of commerce, a sure sign that he had become committed to the region. He had also become a leading figure in regional civic organizations, such as the New Mexico Game Protective Association. After the Armistice, Leopold rejoined the Forest Service as assistant district forester in charge of operations, the second highest position in the unit. In 1921, Chief William B. Greeley offered him a comparable position at the Forest Products Laboratory in Madison, Wisconsin. Leopold declined, but when he was offered the job again three years later, he took it, perhaps believing that he would soon head up the laboratory and thereby advance his career. Or maybe the Midwest was simply in his blood. "My father was Midwestern," his son Carl remarked later; "he spoke with a flat accent."

Among Leopold's enduring farewell bequests to the Southwest was the first federal wilderness reserve surrounding the headwaters of the Gila River. He campaigned hard to protect these lands, and the new reserve was formally dedicated only days after his departure. Wildlands preservation was among Leopold's greatest concerns in the last years of his southwestern sojourn. With Arthur

Carhart, the Forest Service's first full-time landscape architect, he had begun to think and talk about a continent-wide system of designated wilderness areas in the national forests. It was an idea that would gradually take root in American culture, culminating in the Wilderness Act of 1964, a public tribute to John Muir, Leopold, Carhart, and all those who later took up the cause they had championed in the first quarter of the twentieth century. The prominence of wilderness in Leopold's thinking at this time is reflected in the publication in 1925 of no fewer than five major articles by him advocating wilderness preservation. Addressed to varied audiences, the articles offered similar utilitarian yet impassioned arguments for preservation. A sixth, shorter piece of pro-wilderness polemics appeared in an in-house newsletter, the *Service Bulletin*. A seventh was rejected by the *Yale Review*, the literary magazine of Leopold's alma mater.

By the time these fruits of his wilderness thinking were published, however, Leopold had settled into Madison and taken up a new, more sedentary side of Forest Service work. Ten years later, in 1935, he would return to the wilderness cause, joining Robert Marshall and a small cadre of like-minded men to form the Wilderness Society. Marshall beseeched Leopold to accept the presidency. By then, however, Leopold's interest in conservation had taken a different focus, and he declined the offer. He had become the owner of a worn-out eighty-acre farm north of Madison along the Wisconsin River, in a landscape dominated not by big tracts of little-used public land but by thousands of intensively farmed, private smallholds. He was giving himself over, as he wrote in the Wilderness Society's founding year, to "the more important and complex task of mixing a degree of wildness with utility" in the middle, rural landscape. He was at work seeking a lasting, pleasing middle ground between the poles of urban civilization and pristine wilderness.

From Forester to Professor

The Forest Products Laboratory, to which Leopold transferred, was devoted to research and development of such things as wood preservatives and the use of lower-grade trees and mill wastes. Leopold soon found, however, that he had little heart for the business of "pickling ties" and other industry-focused indoor work at the lab. He expressed his true interests in other ways, joining conservation organizations such as the Izaak Walton League of America, making hunting trips to the Ozark Mountains of Missouri and fishing trips to the Quetico-Superior Boundary Waters with his brother Carl and his sons Starker and Luna, and keeping in touch with the Southwest through friends and former associates.

The director of the Forest Products Laboratory, it turned out, did not move on as perhaps Leopold had expected. Feeling trapped and frustrated, Leopold quit the Forest Service for good in 1928. His stature in the conservation community had become so great that he immediately attracted a number of economic opportunities. The one Leopold accepted was the least secure for the long term, but it offered him an opportunity to pursue his ardent passion for game conservation without subjecting his growing family—a fifth child had been born in Wisconsin—to another wrenching move. He became a "consulting forester," funded by the Sporting Arms and Ammunition Manufacturers' Institute to make state-by-state surveys of game in the north-central region of the country, including, all told, the states of Minnesota, Michigan, Wisconsin, Illinois, Iowa, Missouri, Indiana, and Ohio. Leopold enjoyed the work greatly, but, prophet though he may have become in latter-day conservation circles, he did not foresee the stock-market crash that was looming on the horizon.

As the Great Depression worsened, Leopold's survey work withered, and he turned his energies to a book project, *Game Management*. As Leopold then conceived it, game management was the art of producing "crops" of wild animals by manipulating three "factors," their food, cover, and predators. It was a book that drew on all of his experience with game animals, including those as hunter, as well as his extensive travels and interviews with game experts in various states. When it appeared in 1933, it was the first of its kind, and it served for years as the new field's standard text. Soon after the book was published, and after months without income, Leopold secured an appointment to the faculty of the University of Wisconsin as the nation's first professor of game management, with a home in the Department of Agricultural Economics.

From the *Agriculturist* to the *Almanac*

Today, it may be hard to imagine the environmental changes Leopold witnessed in his lifetime. As a boy, he roamed the sloughs and bottoms of the Illinois shore of the Mississippi River, hunting waterfowl. When he returned from prep school in the East one summer by train, he found the wildlands of his boyhood drained and planted to corn. "I liked corn," he later remarked wryly, "but not that much." So obvious had the decline in migratory waterfowl become by the turn of the century that Leopold's father, in the absence of any legal restrictions, imposed bag limits and closed hunting seasons on himself and his sons.

When Leopold was born, much of his native Iowa was unbroken prairie. When he returned to the Midwest as a middle-aged man, he found the country settled up and farmed all out. The ubiquitous and often ill-planned domestication of the landscape made wild things scarce. Compounding the loss of habitat was unregulated shooting. Although there were fewer Americans at that time than there are now, more people lived on the land and derived their sustenance from it. In the first quarter of the twentieth century, most populations of wildlife species, especially those of game animals, were much lower than they would be in the last quarter. To Leopold's discerning and expert eye, the decline in wildlife had reached crisis proportions by the 1930s.

For the Health of the Land is a window into Leopold's outreach work as professor of game management in a college of agriculture at a land grant university in the upper Midwest. This was the task to which Leopold devoted the final decade and a half of his life. This was the task that led him to formulate his much-quoted land ethic. This was the task that inspired and informed *A Sand County Almanac,* his chef d'oeuvre.

The centerpiece of this collection is a series of articles published between 1938 and 1942 in the *Wisconsin Agriculturist and Farmer.* Collected here in part II, "A Landowner's Conservation Almanac," these short essays are mostly how-to pieces. They were part and parcel of Leopold's agricultural extension efforts—practical advice for rural landowners in Leopold's area of special expertise, game management broadly conceived. Although Leopold addressed these pieces explicitly to farmers, he viewed that audience expansively. A farmer, he wrote, was anyone "who determines the plants and animals with which he lives"; it was anyone, that is, who owned or used land and played a role in its destiny.

E. R. McIntyre, editor of the *Wisconsin Agriculturist and Farmer (WAF),* could not have known that this college professor working at the margins of agricultural economics would go on to write the book that would become the bible of today's environmental movement. To us, Leopold's writings are sacrosanct, but to McIntyre, it seems, they were simply journalistic copy to make use of when and how he saw fit. He or his "make-up man" (the person who composed the *WAF* pages) took liberties with Leopold's typescripts, changing titles, adding subheads, and arbitrarily cutting to fit available space. So chagrined was Leopold by one egregious amputation that he had his secretary, Alice Harper, write a letter (dated December 15, 1941) to complain. "Dear Mr. McIntyre," she wrote,

> I should like to call your attention to an error in the printing of Mr. Leopold's December article "Fur Crop in Danger," which appeared in the December 13 [1941] issue of the Wisconsin Agriculturist and Farmer, p. 19.

Directly before the last paragraph of the article, there are almost two paragraphs omitted, thus breaking the continuity. As printed the last part of the story does not have much meaning.

I realize, of course, that it is too late to correct this, but I hope it will not happen again.

In editing this volume, we have taken pains to correct this and other, less draconian instances of cavalier editing. Where possible, we have compared the printed article with Leopold's last-known draft and corrected accordingly. Despite the casual treatment they were given, the items in the series up to mid-1941 were popular enough for McIntyre and his associates to publish as a pamphlet titled *Wildlife Conservation on the Farm.* This republication in pamphlet form gave us as editors another source of comparison to use in putting together as complete and authentic a text as possible. It also supplied the text for one essay that, mysteriously, appeared in the pamphlet but not in the original *WAF* series. After publication of the pamphlet, Leopold continued to write for the *WAF.* When the series abruptly ended, Leopold had completed nine additional essays for it. All are included in this book. Two of them were published by Leopold elsewhere; seven appear here for the first time.

Leopold thought so much of several of his later *WAF* sketches that he reworked them for publication in part I of *A Sand County Almanac.* (We leave the pleasure of discovering which ones they are to those readers for whom the *Almanac* is an old friend.) As early as 1941, Leopold had begun to think of the book that would be his masterpiece, and by 1942, he was drafting chapters for a New York publisher. Thus, work on the *WAF* series overlapped with work on the book that would catapult him to posthumous renown.

From a Crop of Quail to a Swarm of Swallows

Why the *WAF* series ended is not clear from the correspondence record, but it seems that frustration had developed on both sides by 1942. Leopold was frustrated by the indifference to his work shown by McIntyre and his associates and by their delays in publishing it. They in turn appear to have questioned the direction Leopold's writing was taking. At the start, Leopold was working, albeit at the margins, within the field of agronomy. He told farmers how they might grow a secondary or supplemental wild crop of quail, pheasants, and other farm game. Game would never be the main farm crop, but it was a crop nonetheless. As the series progressed, however, Leopold began telling farmers how they might attract songbirds and encourage native wildflowers. He wrote of gazing skyward at geese and woodcocks and improving the aesthetic appearance of farmsteads.

An exchange between Leopold and McIntyre about one sketch, "Cliff Swallows to Order," highlights the apparent rift. On June 30, 1942, Leopold wrote to McIntyre:

> I wish you would tell me frankly whether the wildlife series is unwelcome to your make-up man. I have to spend considerable time on it, and would be glad to avoid the bother if there is any question about the material being wanted. I ask because the swallow story, which I thought was one of the best recent ones, has been in your hands for quite some time, and the season for its appropriateness has just about gone by. . . .

Apparently by coincidence, on that same day McIntyre wrote Leopold a letter:

> I am sorry we have not used the wild life series regularly of late, one reason being that except for the swallow story the rest became a trifle out of season we thought—due to our own delay of course originally, and not yours.
>
> I liked the story . . . of the cliff swallows and went out there

to see them, and talked with the woman of the farm. She said she wished they had never arrived on the premises as they were a nuisance. My wife and I agreed with her upon watching them awhile. Almost too much of a nature event, a deluge where one might enjoy a drink as it were. . . .

Evidently their letters crossed in the mail, and the swallow piece, though later published elsewhere, did not appear in the *WAF*. McIntyre claimed that his readers relished the series, yet he published no more of the wildlife vignettes, and he formally canceled the series a few months later. Ever gracious, Leopold replied, "I entirely understand the situation, and I have guessed as much. I am in no way confusing this with your friendly and appreciative attitude." At that point, however, Leopold not only had begun composing material for the book that would become *A Sand County Almanac* but also had conceived other plans for the *WAF* essays.

From the Pamphlet to the Book

In August 1942, Leopold suggested that the state Conservation Department republish the series "in the form of a pamphlet" for use in the public schools. Nothing came of that proposal, perhaps because the proposed pamphlet seemed too similar to the *WAF* pamphlet—even though Leopold expressly noted that "the series is now extended considerably beyond that included in the pamphlet gotten out by the *Agriculturist*." Not letting the idea die, in a letter dated October 12, 1942, Leopold pitched the project in more detail to Andrew Hopkins, a colleague in the College of Agriculture and editor of its publications. Referring to the *WAF* pamphlet of 1941, he noted that "there are 20 sketches in this reprint. At the present time there are about 35 available." Even so, "the series would doubtless need 'rounding out,'" he continued. But again, nothing came of it.

For the Health of the Land thus fulfills a publishing plan that Leopold himself hatched but was unable to complete. We found a total of forty *WAF* pieces, unpublished and published, unreprinted and reprinted in the 1941 pamphlet or the *Almanac.* And we have "rounded" out the book with a selection of longer essays on the central theme of conserving game — conserving wildlife, more broadly, and preserving land health, more broadly still — in the rural landscape between the poles of urban civilization and untouched wilderness. The essays in part I document Leopold's thinking on this topic before he began writing the series for the *Agriculturist;* those in part III were written during publication of the series or after its termination.

In his correspondence with McIntyre and in his handwritten notes to himself on the matter, Leopold often identified the season, sometimes even the month, for which a given essay was written. Accordingly, we have arranged them in four seasonal sections. Within each, the essays are ordered chronologically, enabling the reader to trace the development of Leopold's style and the evolution — from wild crop to something broader, as already here observed — of the message he sought to convey to landowners.

From Wise Use to Harmony with Land

Early in the twentieth century, Gifford Pinchot, a well-known leader of the Progressive movement in the United States, articulated a philosophy of conservation and molded the nation's first public-lands management agency, the Forest Service, in its image. As a student at the Yale Forest School (founded with Pinchot family funds) and later as a young ranger in the Forest Service (first headed by Pinchot himself), Leopold was thoroughly steeped in Pinchot's utilitarian philosophy of resource conservation. Its motto was "wise use" (now, unfortunately, a phrase appropriated and perverted by the self-styled, anti-environmental Wise Use Movement); its maxim was "the greatest good of the greatest number for the longest time." According to Pinchot, there had been "a fundamental misconception that conservation means nothing but the husbanding of resources for future generations. There could be no more serious mistake. The first great fact about conservation is that it stands for development." What distinguished conservation from resource rapine, in Pinchot's view, was efficiency and equity. Informed by science, foresters and other stewards could ensure that nature's renewable resources were not wasted or exploited in a way that diminished their regeneration. Public ownership of natural resources could ensure that profits from their development would be democratically shared.

Pinchot's strident tone was doubtless meant to head off any tendency to conflate resource conservation with nature preservation, then the principal alternative philosophy of public-lands management. Nature preservation was vigorously championed by Pinchot's older contemporary John Muir. Most preservationists thought—and the bolder among them, such as Muir, actually said—that Nature (with a capital *N*) was sacred. To conserve nature was to protect it from human interference. Whereas resource conservation implied "wise use," or well-managed exploitation of natural resources, nature preservation implied the setting aside—Pinchot called it "locking up"—of land and the prohibition of all extractive or consumptive uses, especially in the country's most majestic natural places. Just as the national forests (where extraction of natural resources is allowed and encouraged) are a legacy of the resource-conservation paradigm, the national parks and monuments (where no extraction is allowed) are a legacy of the nature-preservation paradigm.

Leopold's advocacy of a system of wilderness areas in the national forests was thus seen as apostasy by people such as his first boss back in the Southwest, John Guthrie. "Forestry is not aesthetics, is not 'natural areas,' nor wilderness per se," Guthrie proclaimed in a letter to Leopold, "but the putting to use, and *commercial* use at that, of all the resources of the country." As he was then still in the employ of the Forest Service, Leopold carefully cast his essentially preservationist program in the rhetoric of resource conservation. Being essentially sacred spaces, the national parks were appropriately off-limits to hunting. But wilderness preserves in the national forests, Leopold argued, should still be understood as working, productive lands. Recreation (mainly hunting) was the "highest use" of those hitherto unexploited areas in the national forests that were too poor, too remote, or too rugged for profitable logging or grazing. Game, and such intangible goods as sport and primitive forms of travel, thus became for Leopold the forest and range "products" of designated wilderness areas in the national forests. It was an argument very much in the Pinchot spirit. But between the lines, one senses—unstated yet unmistakably present—the conviction that wildlands have intrinsic value and deserve preservation for their own sake.

Leopold is therefore often seen as a conservationist who began his career in the Pinchot camp and gradually came over to the Muir camp. The true story, however, is more complex. In the Southwest, the internecine conservation battle lines were drawn between efficient resource exploitation and wilderness preservation. But the Midwest to which Leopold returned in the 1920s lacked extensive public lands and national park grandeur. The great pineries of Michigan and Wisconsin had long been felled. The great prairies of Illinois and Iowa had long been plowed and planted to annual grains. Except for the Quetico-Superior Boundary Waters, little wilderness was left to preserve. The landscape had been largely carved up into privately owned smallholds. How could a conservationist apply either the Pinchot paradigm or the Muir paradigm in this kind of country?

Conventional agronomy in the Midwest was devoted to the efficient exploitation of land, and it was to the conservation challenge posed by such exploitation that Leopold turned his attention and his sensibilities soon after his arrival there. Something was plainly missing in the hardworking farm landscape that surrounded his new hometown. Big wilderness was out of the question; so was big-scale forestry. What did seem possible, Leopold decided, was the integration of a degree of wildness into the working landscape mosaic of cultivated fields, pastures, woodlots, and wetlands. But what to call this third conservation paradigm, this middle ground between unsullied wilderness and unrestrained exploitation? In time, Leopold would give it various names: a "harmony between men and land," "a mixture of beauty and utility," "the principle of wholeness in the farm landscape," "a state of health in the land-organism," and finally and most simply, "land-health."

From Wilderness Game to Farm Game

The unifying thread through Leopold's intellectual odyssey, from resourcism (Pinchot's philosophy) toward preservationism (Muir's philosophy) to his own harmony-with-nature philosophy, is Leopold's interest in game. At the end of the nineteenth century, North America's most magnificent fauna had been reduced to near extinction (indeed, to actual extinction in the case of the passenger pigeon) by unregulated market hunting. The people who took the lead in responding to this particular tragedy of the commons were sport hunters. Relatively affluent and often well connected, they succeeded in getting closed-season and bag-limit legislation enacted—thus saving their sport from drying up and permitting a gradual recovery of game populations. At the end of the twentieth century, with the vocal presence of People for the Ethical Treatment of Animals and similar groups, it is sometimes forgotten that sport hunters were once the unquestioned good guys of conservation. The bad guys were greedy market hunters, who killed animals by the thousands to sell for food and ornamentation. To Leopold, as he made the transition from the Southwest to the Midwest and from a civil-service to an academic career, conservation was practically synonymous with making sure that sport hunters would forever enjoy plentiful game.

Pinchot's philosophy, as paradigmatically applied to forestry, provided the youthful Leopold with a model for scientific game management. Midway through his tenure in the Southwest, Leopold cast his first formal game-management essay in just this Pinchovian paradigm of conservation: a game census was to game management what reconnaissance was to forestry; law enforcement against poaching was analogous to fire control; breeding stock, to seed trees; license fees, to stumpage rates; bag limits and closed seasons, to limitation of cut; and the game farm was to game management what the tree nursery was to forestry. Indeed, so enthralled was Leopold at this point with Pinchot's philosophy that he could essay to debunk the "popular wilderness fallacy"— the "fallacy" that with the rapid reduction of wilderness to fair farms, ranches, and productive forests, game would grow scarcer. As his ecological understanding broadened, Leopold came to repent such follies and to envision wilderness preserves in the national forests as enclaves that were vitally needed if game was to survive.

When he moved to the Midwest, Leopold shifted his focus from the "wilderness game" of the southwestern forests and rangelands to the "farm game" that inhabited Wisconsin's dairy lands. With the exception of the white-tailed deer, the Midwest's wilderness game was long gone, and he was saddened by its loss. But even worse for Leopold, many once abundant farm-game animals were suffering marked declines, and few people seemed to know what to do about it. The recovery of farm game was an obvious conservation challenge, and Leopold took it on as his own.

From Game to Wildlife

The essays in part I of this volume, "Conserving Rural Wildlife," all written in the 1930s, are the fruits of Leopold's deepening reinhabitation of the Midwest and the shift in his career from forestry to the new profession he pioneered, game management. They help round out the volume. In them, we can see Leopold evolving from a narrow preoccupation with game to a wider concern for wildlife, irrespective of its value as a sportsman's target.

The first essay, "Game Management: A New Field for Science," appearing here for the first time, was written for *Scientific American* but never published there. It provides a window into Leopold's game-survey research and documents his efforts to turn an amateur undertaking into a new professional field.

The second, "Helping Ourselves: Being the Adventures of a Farmer and a Sportsman Who Produced Their Own Shooting Ground," is a charming account of a personal experiment with farmer-sportsman cooperatives in the production and harvest of wild crops. The "set up" Leopold describes was known as the Riley Game Cooperative, and it eventually became a site for wildlife research by his graduate students. An essay in part III continues the Riley story.

The third, "The Wisconsin River Marshes," focuses on game but not on farmers. In it, Leopold envisions the prospect of a small-scale semi-wilderness reserve strung out along the Wisconsin River in the region that Leopold would later immortalize as the sand counties.

The fourth, "Coon Valley: An Adventure in Cooperative Conservation," is a mirror image of the third, focusing on farmers but not on game. Leopold deals here with a problem that concerned him in the Southwest and that also plagued the Midwest—soil erosion.

The fifth, "Farm Game Management in Silesia," details what Leopold learned about game propagation on farmlands in Germany, which he visited with a contingent of American foresters in 1935. Leopold was deeply disturbed by much of what he saw in Germany, particularly industrial forestry and artificially inflated deer herds, but as this paper indicates, he was favorably impressed by some aspects of German farm-game management.

The sixth and final essay in part I, "Be Your Own Emperor," is, like the first, also published here for the first time. It provides an overview of Leopold's efforts to establish the new profession of game management in the Midwest on the eve of his pivotal *WAF* series. Leopold clearly heralds here the extension of scientific management to all kinds of wildlife—even to wildflowers. Notably, he expresses a tolerance for predators and defends them against vilification as "vermin," a prejudice of which he was himself once guilty.

From Wildlife to Land Health

Part III, "Conservation and Land Health," consists of seven essays. By including them, we intend to do more than merely fulfill Leopold's plan to round out the volume. The material in parts I and II leads up to the philosophical issues that the author more directly confronts in this final part of the book.

Part III opens with "The Farmer as a Conservationist," which begins with Leopold's succinct definition of conservation as a "harmony between men and land." This wonderful idyll sums up his thinking about rural-lands conservation as it had begun to galvanize in his extension work with farmers and in his writing for the *WAF*. Farm game is a concern of this paper, but only in passing. The central concern is the maintenance of *native* plants and animals on the farm and an appreciation of the human benefits of such maintenance. He suggests that "wise use" and wilderness preservation are just different points on a gradient of "restraint." His own novel philosophy of conservation is more proactive: "a positive exercise of skill and insight, not merely a negative exercise of abstinence or caution." Also mentioned here for the first time is the idea that land, no less than the human beings who live on it and by it, can be in a state of "good health."

The next essay in part III, "History of the Riley Game Cooperative, 1931–1939," was written in the spirit of "Be Your Own Emperor"—more a candid scientific assessment than inspirational testimony. It should be read as an antidote to the more upbeat "Helping Ourselves."

Among the most oft-quoted of Leopold's many turns of phrase is found in his book *Round River:* "To keep every cog and wheel is the first precaution of intelligent tinkering." Published here for the first time, "Planning for Wildlife" expands on that maxim and further develops the concept of land health. So ingrained in Leopold's thinking was his new conservation paradigm that in this essay he reconceived the principal rationale for wilderness preservation. "Every region should retain representative samples of its original or wilderness condition," he wrote, "to serve science as a sample of normality" in land physiology and metabolism. Such samples would be "control" areas, benchmarks, enabling society to measure its success in living on the land and deriving a livelihood from it without degrading its health.

The fourth paper in part III is the keystone of the book. Published here for the first time, "Biotic Land-Use" links land health with stability and stability with diversity. In it, we find yet another of Leopold's attempts to define clearly his novel philosophy of conservation: "The term 'land' includes soils, water systems, and wild and tame plants and animals. Conservation is the attempt to understand the interactions of these components of land, and to guide their collective behavior under human dominance." Without using the term, Leopold here concisely articulates the core approach to land stewardship now called ecosystem management. First, he identifies a suite of management "technologies": agronomy, erosion control, flood control, pasture management, forestry, and wildlife management. Practiced separately, he notes, they are mutually contradictory. For them to be mutually consistent and reinforcing requires more than just coordination and integration. What is required is a common goal or purpose at which all aim: "the health of the land as a whole." This purpose does not cancel the separate goals of each—crop production in the case of agronomy, timber production in the case of forestry, and game production in the case of wildlife management. Rather, it subordinates each of these goals to the common one.

Leopold's shift from an exclusive focus on conserving game to a much broader conservation scope is nicely illustrated in "What Is a Weed?" In this gem of irony and understatement, his concern for the native prairie flora is manifest. But Leopold did not contract a case of floristic xenophobia—he did not spurn all non-native species—as many conservation biologists have today. He shows a tolerance, even a fondness, for some "immigrant" plants. Perhaps they too, he suggests, can play a small physiological role, and play it well, in a healthy land organism.

Three years after the *WAF* series ended, Leopold assessed "The Outlook for Farm Wildlife" for his fellow professionals, and it was not optimistic. The increased mechanization of the farmstead and the irresistibility of the industrial model for agriculture—so vigorously promoted by most of his agricultural colleagues at the university—were overwhelming Leopold's best efforts to foster a mix of wild with tame crops and beauty with utility.

The final essay in this book, "The Land-Health Concept and Conservation," is published here for the first time. The concept that had gradually emerged in Leopold's conservation thinking here takes center stage. Leopold defines land health as "the capacity for self-renewal in the biota." He closely connects it with another concept now much discussed in conservation circles— ecological integrity, the presence of the full complement of the native components of biotic communities in their characteristic numbers. One can feel in this piece the same mood and tone of Leopold's most famous essay, "The Land Ethic." And it is as fitting a conclusion of this book as "The Land Ethic" is of *A Sand County Almanac.* In it, we find the central themes of farmers and wildlife in the context of Leopold's conservation ideal—an optimal mix of the wild and the tame, of beauty and utility, in a humanized landscape epitomized by the Midwest.

From Private Property to Community Responsibility

By the time Leopold had written the essays in this book, one further shift had occurred in his understanding of conservation, and it was as foundational and far-reaching as any other.

When he began writing seriously about game management, Leopold knew that it was not game that required management but the habitat where game lived. To manage game was to manage land. Soon, though, he sensed that the chief obstacle to sound land management was not so much the difficulty of restoring habitat as the difficulty of redirecting the attitudes and values of landowners and land users. Before sound ideas could take root, unsound ways of thought needed pruning. Conventional wisdom needed challenging, and Leopold set out to challenge it—the "wisdom" of killing predators without cause and study, of draining wetlands dry, of practicing "slick and clean" farming, of removing fencerows and windbreaks, of straightening streams, even of showing disdain toward ragweed and foxtails. He was particularly hard hitting when bad advice flowed from universities and agricultural extension agencies, as it often did. Leopold also turned his critical gaze on conservation bureaus, which too often addressed not the etiology of land sickness but merely the outward symptoms. They built check dams to catch eroded soil rather than keeping the soil well vegetated; they built fish hatcheries to restore fisheries rather than addressing the underlying causes of declining fish populations.

For years, Leopold pondered the motives of landowners, wondering how conservation policies might lead toward better land practices. At first, he placed hope in economic incentives, including farmer-sportsman arrangements such as the one tried out at Riley, but by the time he began the *WAF* series, he knew that more was needed. Good land use was an art, and it required great skill and careful study by dedicated land managers. Landowners had, however, to put their hearts and souls into the work as well as their minds. They had to see beauty in native plants and animals and to take satisfaction in owning land resplendent in diversity and health. Above all, they had to see how they and their lands helped form an overall land community. To live responsibly in that community, a landowner should pay attention to its needs and limitations and should shoulder a fair share of community responsibilities. At its base, conservation was fundamentally a moral issue, however important science had become. It was a matter of loving the land and feeling tied to it, a matter of standing tall as a community member and working for the good of the whole.

Two years after the *WAF* series ended, Leopold distilled its messages into a short summary. His summary centered not on game nor on the land, nor even directly on land management. It centered instead on "the farmer," that all-important community member who directed the land and made decisions about its plant and animal members. "The farmer," Leopold wrote,

> should know the original as well as the introduced components of his land, and take pride in retaining at least a sample of all of them. In addition to healthy soil, crops, and livestock, he should know and feel a pride in a healthy sample of marsh, wood lot, pond, stream, bog, or roadside prairie. In addition to being a conscious citizen of his political, social, and economic community, he should be a conscious citizen of his watershed, his migratory bird flyway, his biotic zone. Wild crops as well as tame crops should be part of his scheme of farm management. He should hate no native animal or plant, but only excess or extinction in any of them.

When in "The Farmer as a Conservationist" Leopold wrote that "the landscape of any farm is the owner's portrait of himself," he meant not just a portrait of the owner as farm operator but a portrait of the owner's inner moral character, his sense of aesthetics, his willingness to step forward and uphold the land community. It was a portrait of his ecological wisdom, his artistry, and his ethics.

Although Leopold always respected private property, in the end he came to understand that landownership inevitably meant community membership. Along with the rights of ownership came duties to the community, duties to leave room for wildlife, to keep soil in place, to leave hydrologic flows sufficiently natural, and to ensure in other ways that the owner's tract of land contributed to the health of the larger landscape. One of Leopold's aims in the essays included here was to craft a clearer sense of what community membership meant—to explain the individual's rightful role and to set forth the owner's fair share of communal burdens. Reciprocally, he also argued that communities bear a responsibility to help support the individual—especially the hard-pressed farmer who must shoulder such communal burdens.

What Leopold sought in his writings on conservation of private lands, then, was a transformation of the heart and soul of the landowner as much as a thoughtful transformation of the rural landscape guided by ecology. And he approached that task with hope and zeal. The challenge, he knew, was great, but the payoff was even greater. A healthy land community would mean a healthier, happier, more satisfying life for its human members. It would mean living in a land of greater beauty. It would mean more intimate contact with the abiding forces of nature.

Paradoxically, then, the rounded-out book that Leopold had in mind during the darkest moments of World War II looked toward the advent of an ecologically informed environmental ethic and aesthetic. And it is to that same end that this book is offered today. Some of Leopold's land-management recommendations inevitably require updating, given the changed circumstances since Leopold wrote and given the subsequent development of ecological science. But Aldo Leopold had a dream, and it remains as inspiring as ever. It was a dream in which conservation—the conservation of the entire land community—was fostered not just in national parks, wilderness areas, and other public domains but in every corner of every landscape. It was a dream in which conservation became a down-to-earth endeavor for all users of the land—for all farmers, as he called them; an endeavor pursued in every watershed, in every forest and field, on the back forty, even in the backyard—in every place where people lived on the land and helped guide its unfolding self-renewal.

PART I

Conserving Rural Wildlife

Game Management: A New Field for Science

In this brief essay, written for Scientific American *in 1932 but never published, Leopold introduces many of the ideas that would thereafter appear conspicuously in his work, particularly the idea that game was best promoted by providing good habitat and letting nature do the rest. It was an idea familiar in other countries and cultures but not yet known in the United States and nowhere practiced with the ecological rigor that Leopold would bring to bear. In this essay, Leopold was already criticizing "modern" methods of "slick and clean" agriculture, which stripped farm fields of wildlife cover and food, and calling for wide-ranging, practical experiments to determine which land treatments did and did not promote game. When he wrote this essay, Leopold apparently thought that "slight modifications" in farming practices would suffice to yield abundant farm game and that producing game could be "an economic move for the farmer." He later realized that the needed changes were far greater and the economics less appealing.*

THE FLOOD of science which has inundated the affairs of everyday life has, like any other swift current, exhibited certain backwashes or eddies. For a long time one of these "slow spots" was agriculture—the art of producing crops of domesticated plants and animals. There is still another "slow spot" which has just begun to feel the impetuous onset of the scientific method. This is game management—the art of producing crops of *wild* animals for recreational use.

The "best" game crop is of course that which has produced itself, without human aid or interference. It is increasingly evident, however, that in settled countries the entire lack of human aid means the ultimate obliteration of wild game, and also the obliteration of the various field sports and recreations involved in its pursuit.

Some kinds of game can, of course, be confined and artificially propagated, but the costs per head are high, while the product, even after release to the fields and woods, lacks "that something" which the American sportsman demands in his quarry. Hence artificial propagation, while an invaluable source of "seed stock," is not by itself a sufficient answer to the question of game shortage.

Wild animals, under ideally favorable conditions, increase with incredible rapidity. Is there not some way in which a part of the tremendous "breeding potential" can be realized—some way in which, by increasing the wild survival, an abundant annual crop of wild game can be restored?

To search for an answer to this question, some of the industries affected by the decline in the game supply have financed a "Game Survey." A modern Solomon might say "of the making of surveys there is no end," but label it what you will, this Survey is an attempt to appraise the chances for game restoration in America. It has so far examined a block of eight states in the upper Mississippi Valley. A report of its findings has been recently published.*

The fundamental reason for game scarcity in this region, the report says, is a new entity without a name. A name for it has been borrowed from the physicist and the electrical engineer. It is "environmental resistance." The meaning is this: "Slick and clean" agriculture has removed the game food and cover, and thus increased the resistance which the environment offers to natural increase.

Game Survey of the North Central States (Washington, D.C.: American Game Association, 1931).

Hence a lesser fraction of the breeding potential is realized, hence game is decreasing.

A sample of the evidence on which the findings of the Game Survey are based is the analysis of quail "density" (population per unit area) in various states. A quail census was made of nearly 400 sample ranges, each about the size of an ordinary farm, and the density of each computed by dividing the quail population by the area. A map of these densities shows low abundance to prevail in the prairie farming regions, whether shot or unshot, and especially where farming custom has decreed the removal of hedge fences and fencerows which serve as quail cover. The map shows high densities to prevail in the semifarmed Ozark foothills, and along the riverbreaks of the prairie states, where brushy woodlots, draws, and fencerows are still found, but where grain is also available for winter food. Low densities, on the other hand, prevail in the high Ozarks, where there is endless cover but little grain.

The conclusion indicated is that environmental resistance to quail increase is least where agriculture methods happen to provide food and cover in close juxtaposition, but that the resistance rises where either factor gets "out of balance" with the other.

The same "law" is apparent from an historical study of food and cover changes on a single farm through a long period of time. The Report cites the Phil M. Smith farm in Missouri (see diagram) which, in 1923, supported over 200 quail. At that time it was a "backward" farm, with plenty of brushy draws, hedges, and an ungrazed woodlot. For the next five years its owner, who had just graduated from an agricultural college, subjected it to the process of "modernization." As the livestock, brush-grubbing, and fence-clearing on this farm increased, the quail decreased. In 1929 less than half the original quail population remained. The Report says:

> There has undoubtedly been a large increase in the sale value of the farm, due to the enhanced working capital of livestock, fertility, and pasture area.

The question is, however, whether this enhanced value of the farm as a productive agricultural unit could not have been attained without so heavy a sacrifice of its game-producing capacity.

There can be no absolute yes or no answer to this question. The answer is a matter of degree.

The question is statewide, because thousands of farmers are doing, or will eventually do, the same thing. A powerful and extremely effective machinery is maintained in each county at governmental expense to hasten the process, and to show the farmer how. Could not this same machinery show him how to conserve at least a part of his game-producing capacity, if he cares to do so for either pleasure or profit?

To illustrate concretely what is meant by "conserving game-producing capacity" on an improved farm, the map depicts an imaginary reconstruction of the quail coverts on the Smith farm, and the hypothetical response in quail population. The intent is to show how "concentrated" coverts can be squeezed into odd corners without sacrifice of valuable acreage. Whether agricultural experts would approve this particular reconstruction is not known. Probably not. The major plea is that they start experiments on this question, and tell the farmers what particular measures for the benefit of game *would* meet with their approval, and what response in game might be expected.

The technical process of adapting agriculture and game production to each other—this art of raising game as a wild by-product of the land—the Report calls "Game Management." It offers an almost virgin field for the practical application of biological science. Its techniques are just beginning to be developed. To hasten their development, and start the training of a professional class competent to apply them, the ammunition industry has set up a series of research fellowships at several universities or agricultural colleges. One in Minnesota is studying the management of ruffed grouse, at Wisconsin the management of bobwhite quail, at Michigan of Hungarian partridge, at Arizona of Gambel's quail. Under

Cover and Crops in 1923

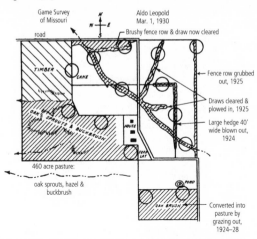

Game Survey of Missouri

Aldo Leopold
Mar. 1, 1930

road

Brushy fence row & draw now cleared

TIMBER

CANE

OAK SPROUTS & BUCKBRUSH

HOUSE

FEED LOT

← Fence row grubbed out, 1925

Draws cleared & plowed in, 1925

Large hedge 40' wide blown out, 1924

460 acre pasture:
oak sprouts, hazel & buckbrush

POND

OAK BRUSH

Converted into pasture by grazing out, 1924–28

Cover and Crops in 1929

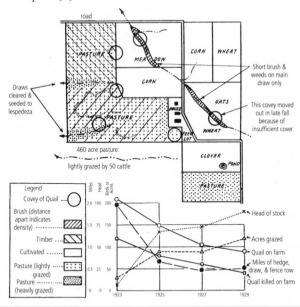

road

PASTURE

MEADOW

CORN

CORN WHEAT

HOUSE

OATS

FEED LOT

WHEAT

Draws cleared & seeded to lespedeza

Short brush & weeds on main draw only

This covey moved out in late fall because of insufficient cover

460 acre pasture:
lightly grazed by 50 cattle

CLOVER   POND

PASTURE

Legend

- Covey of Quail
- Brush (distance apart indicates density)
- Timber
- Cultivated
- Pasture (lightly grazed)
- Pasture (heavily grazed)

Miles / Head / Birds or Acres

Head of stock
Acres grazed
Quail on farm
Miles of hedge, draw, & fence row
Quail killed on farm

1923 1925 1927 1929

FIGURE 1: Effect of agricultural improvements on quail (Phil M. Smith farm—280 acres, 3 miles west of Williamsburg in Callaway County, Missouri)

Cover and Crops as They Might Be

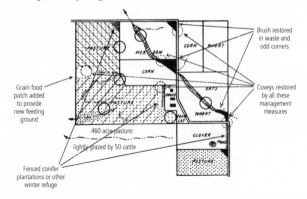

FIGURE 1 *(continued)*: Effect of agricultural improvements on quail (Phil M. Smith farm—280 acres, 3 miles west of Williamsburg in Callaway County, Missouri)

the eye of the agricultural authorities, and with the advisory guidance of the U.S. Biological Survey, these "Game Fellows" are amassing a body of skill on how to raise wild game on modern farms and in modern forests. The keynote of the whole venture is to find out what slight modifications in methods of managing the primary crop will decrease the environmental resistance to the increase of game.

That agricultural crops are overproduced is now universally admitted. It would seem to follow that the dedication of the poorer parts of thousands of farms to valuable game crops would be an economic move for the farmer, and a substantial answer to the unsolved question of game conservation. Game is the only land crop in which there is no present or prospective overproduction. An unlimited market for hunting privileges exists in the form of five million hunters, annually licensed by our various states to hunt a game crop which in many regions is gradually passing out of existence.

Helping Ourselves: Being the Adventures of a Farmer and a Sportsman Who Produced Their Own Shooting Ground

In this charming piece published in 1934 in Field and Stream, *Leopold recounts how he and a local farmer, Reuben Paulson, organized the Riley Game Cooperative. Riley was just the kind of practical experiment in private-lands management that Leopold advocated in the previous essay. We see here Leopold's faith in citizen action, his attention to economic realities, and his infectious "incurable interest in all wild things." Although Paulson is named as "the junior author," the essay is written in Leopold's style from beginning to end.*

THE SENIOR author of this narrative is a sportsman who had grown tired of asking suspicious farmers for permission to hunt, hike, or train dogs on gameless farms. The junior author is a farmer who had grown tired of spending his Sundays ejecting miscellaneous unpermitted "rabbit-hunters" from his quail coverts.

Like other outdoorsmen, both of us had listened patiently to the fair words of the prophets of conservation, predicting the early restoration of outdoor Wisconsin. We both had noticed, though, that as prophecies became thicker and thicker open seasons for hunting became shorter and shorter, and wild life scarcer and scarcer.

Three years ago, when we first met, to flush a rabbit was the biggest adventure one might hope to fall upon in a day's hike on the Paulson farm. One snowy Sunday, when we were bemoaning this scarcity of living things on the land, there came to us jointly a flickering recollection of that first theorem of social justice: The Lord helps those who help themselves. Whereupon was born the "Riley Game Cooperative."

Riley, be it known, is a flag-station and a post-office near the

Paulson farm. This definition of Riley is meticulously and literally correct.

The term "game cooperative" was not quite so accurate. It was a "cooperative," all right, with one farmer and one sportsman constituting its then membership. But it was more than "game," both of us contributing to the enterprise an incurable interest in all wild things, great and small, shootable and non-shootable. However, we both had an eye cocked on the future, and decided to title only the main issue.

Paulson gathered unto himself six contiguous neighbors. Leopold gathered up five Madison sportsmen, all mutual friends and of the sort whose game pockets contain no quail feathers in pheasant season. Then we moved that the nominations be closed. The idea is that any enduring relationship between sportsmen and farmers must be based on personal confidence, and nobody can have that if the crowd is so large as to need identification tickets. We of the Cooperative can name any other member across the marsh by noting the decrepitude of his particular hunting coat, or by watching the gait or ear-floppings of his particular dog.

Now it so happens that in that same winter of our discontent, when the first theorem of social justice was revealed to us, some senator or assemblyman likewise saw the burning bush. We admit that legislators seldom do this, either in Wisconsin or elsewhere, but this one did. There emerged, as out of a cloud, all duly enacted, the "Wisconsin Shooting Preserve Law," which declared that citizens who owned or controlled land and planted pheasants thereon might shoot, when duly licensed, three-quarters of the number planted, during an all-fall open season, provided there be affixed to the leg of each pheasant so shot a non-reusable metal tag, to be issued by the Conservation Commission, etc. Furthermore, the law prohibited trespass by other citizens on the premises so licensed.

The law specifies pheasants, because these can be raised artificially; and when they are counted out of the coop by the local game warden, the state knows what three-quarters is. The state gets the

other quarter "on the hoof," as a private donation, to chalk up to
the credit of its restocking program.

We of the Cooperative are no more interested in pheasants than
in other game, and still less in shooting pheasants recently let out
of a coop. Be it noted, however, that the new law restricts shooting
to three-quarters of the number released, not to three-quarters of
the identical birds released. We saw in this a chance to build up a
wild population and to do our shooting on these wild birds, releas-
ing sufficient tame ones to satisfy the requirements of the law.

Therefore, we took out a shooting-preserve license, posted the
seven member farms, and released twenty-five pen-raised pheasants as a starter. None of us shot them, or wanted to, but we all had
a lot of fun that first winter maintaining feeding stations "for the
succour of said beasts," and, to be honest, for the purpose of holding them on our grounds. It was a mild winter, and these "tame"
pheasants soon grew too big and wild to be in need of much "succour."

It was, however, the patronage extended to our feeding stations
by non-shootable game which made them fun. Paulson had planted
soy-beans under his silage corn. An aftermath of these beans had
matured after the corn harvest. At the very first heavy snow these
soy-beans drew, out of nowhere, a pack of forty big, husky prairie
chickens. No chicken had been seen on the Paulson farm for a
decade.

Likewise out of nowhere came a covey of quail. They tried to
establish legal residence at one of the pheasant stations, but it was
soon evident, from the lawsuits recorded in the snow, that the
pheasants disputed their emigration papers, and not always by
peaceable means. So we promptly erected an additional station for
the quail, and henceforth each species stayed in its own bailiwick.

Before the winter was over a second quail covey, doubtless
starved out of some near-by farm, appeared and waxed fat at our
expense. Only it shouldn't be called expense—none of us for years
had so enjoyed our winter Sundays. As for rabbits, every one

within a mile of our boundaries promptly applied for membership in the corn supply of the Riley Cooperative, and when winter was over they stayed to set up housekeeping.

It was a pleasant thing that first spring, as we strolled over these formerly gameless farms, to hear quail whistling in every fencerow and pheasant cocks crowing all over the Sugar Creek marsh. We estimated that our first six months of operation had netted us a respectable pheasant population (some strayed to the "public domain," as predicted by the law) plus an unearned increment of thirty quail, plus bunnies *ad infinitum.* Our chickens left us for parts unknown after the last snow had melted, but we knew that they would be back.

It was now time to do our stuff under the preserve law. Buying grown pheasants at $2.50 each was too expensive; so we bought 150 eggs, and Mrs. Paulson hatched them under hens. When the game

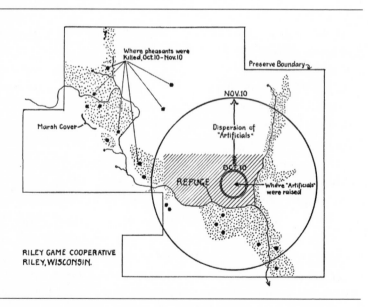

FIGURE 1: How a refuge confines shooting on an artificially stocked preserve to wild birds

warden came around in August, he counted 70 half-grown birds, which had the free run of the orchard but returned to roost in the brooder-coop with their foster-mothers. This count entitled us to 53 shooting tags (three-quarters of 70), plus those unused last year. These tags are equally distributed among all our members who care to shoot, including farm members.

At this point we hear our sporting readers emit a loud snort at the prospect of shooting these half-tame "artificials." Just a minute, please. These tame hen-raised birds were all headquartered near the farmhouse, around which we blocked off an eighty-acre refuge on which no shooting is allowed. Outside this refuge, ever since the corn was cut last fall, we had been training our dogs on several coveys of big wild birds, the progeny of last year's plantings. It was these wild birds that we hunted when our season opened in October. By late fall the "artificials" had gone wild and spread, by slow degrees, off the refuge, and we probably shot some of them, but at no time had we either the desire or the opportunity to shoot an immature or tame pheasant. Our refuge automatically prevents it. The sketch map on the opposite page shows how this refuge works.

At this writing, our stock of game compares with the three preceding years as follows:

	Nov. 1930	Nov. 1931	Nov. 1932	Nov. 1933
	(Before Management)			
Pheasants	None	25	100	125
Quail	15	30	90	150
Prairie chickens (winter only)	None	40	60	30
Rabbits (estimate)	50	100	125	125
Total head	65	195	375	430

Our main loss of pheasants was due to birds flying off the preserve when shot at. In 1932 we killed twenty-five, but doubtless lost twice that number from scattering. Last year we seem to have partly overcome this by more refuges, more feeding stations, and postponing the shooting till November and December. At this season the outside range is bare and unfed, and hence less attractive to the birds.

Feeding is our main bid for permanent residence. We have had no luck with food patches because our farmers turn their hogs loose in the fall, and they make short work of any grain left standing. However, a dozen lean-to shelters, built of poles and thatched with marsh hay and each containing a hopper or a wire crib, are on duty all winter as feeding stations. A hundred-yard circle around each feeding shelter is closed to shooting.

We settled, by a tribal moot held under Paulson's oak tree, the financial relations of town and farm members. All costs are to be shared equally. When a farm member or his wife raises pheasants, he or she is to be paid, by the town members jointly, for each pheasant counted out by the warden, half the game-farm price. When a farm member leaves grain for a feeding station, he is to be paid, by the town members jointly, half its market value. All this proceeds on the theory that about half the "keep" of a game crop is the land on which it ranges, which land the farm member is furnishing free.

All shooting privileges are pooled and equally divided among those members who wish to shoot. Some of the farmers do not care to shoot, but they are none the less enthusiastic members of the Cooperative, by reason of the pleasure they derive from seeing the game, the insect-control services rendered them by the game, and the protection from irresponsible trespassers derived from the posting of the preserve. Nor should it be forgotten that the state is likewise a beneficiary. Its tangible reward is the $10 we pay the Conservation Department for a preserve license, plus at least fifty unshot pheasants, worth $2 apiece, which have spread over the country. Its intangible reward is the manifold increase we have brought about in the quail and songbirds which use our feeding stations and coverts. We of the Cooperative have likewise decreased by twelve men that growing army of shooters who have no place to shoot and nothing to shoot at.

So far we have found it unnecessary to control any predators except cats. We encourage local trappers to trim down our foxes and minks, but select those individuals whom we think can be trusted to let feathers alone. We have horned owls and red-tailed hawks, but there is no evidence that they are getting birds. Our abundant rabbits seem to serve as buffers. Skunks are plentiful, but their droppings in summer are a solid mass of June-bug wings. Hence our farmers, whose pastures suffer from the grubs, want the skunks left undisturbed until actual nest-robbing becomes evident.

As our bird stock builds up, the predator situation may change, but we will molest no predators until they molest us. If and when we take measures against them, it will be with genuine regret. It is pleasant to hear the owls hoot as we tramp carward from a day's hunting, and on winter Sundays the fox-tracks in the snow add interest to our rounds of the feeding stations.

During the years we have shot over our preserve, the total of pheasants killed has never reached half of the number we were entitled to shoot. The birds killed were nearly all cocks, although our own rules allow one hen for each two cocks. Nobody has killed more than one bird in a day, although perfectly free to shoot his whole quota. All this goes to prove the astonishing conservatism of the "blood-thirsty" hunter once he begins to feel a proprietary interest in the future game crop of a particular piece of land.

To kill one's first cock on one's own grounds is a memorable experience. This cock, according to the trail unraveled by the spaniel pup and the food later found in his crop, had breakfasted in the cornfield and then repaired to a willow-bush in the marsh to repose in the mild October sun. Was this a "tame" pheasant? Not very. He hurried out of that willow-bush, in all his bronze and violet glory, like some indignant Genghis Khan disturbed at his nap, the marsh resounding to his profane cacklings and the great tattoo of his broad, strong wings. It took both barrels before he collapsed into the marsh, and when the pup proudly emerged from the grass bearing his limp and shining burden you could see only his eyes and paws. The rest was all pheasant. One of these big, wild scarletjowled white-collared cocks is enough flavoring for any Sunday afield.

We want other farmer-sportsman groups to set up game production around our boundaries. This cooperative idea has all outdoors to spread in. We have not so far been accused of monopolizing shooting privileges, but we will be; and when we are accused, this will be our answer:

"There are 12,000,000 acres of farm land in southern Wisconsin, capable of carrying at least 6,000,000 game birds, of which 2,000,000 to 3,000,000 could safely be shot annually, if and when they are brought into existence. There are possibly 100,000 shooters on this area, which means that each could shoot 20 to 30 birds per year, if and when the farmers and sportsmen will get together, under shooting-preserve licenses or otherwise, and provide food and cover so that the birds can multiply. The state cannot provide food and cover. Farmers can. The Lord helps those who help themselves."

The Wisconsin River Marshes

In this essay, published in the National Waltonian *in 1934, Leopold describes another way in which landowners might organize to achieve conservation goals that are beyond the reach of any single owner acting alone. Moderate, ecologically sensitive uses of marshlands, he tells us, can be good for both people and wildlife. But destruction and decline ultimately come when "the march of empire" alters the land too much. Leopold's farsightedness shows up here in his calls for plugging drainage ditches, reflooding marshes, and encouraging farmers to form compact communities. We also see his appreciation of the vastness of human ignorance about the land, illustrated by his long list of questions that need answering just to "reflood a marsh and let the wildlife take it."*

AN INSTRUCTIVE example of how economic forces and public sentiment play basketball with conservation is found in the marshland district of central Wisconsin.

These marshes are strung out on both sides of the Wisconsin River and cover about half a million acres in five counties. They consist of peat-filled basins which represent the deeper parts of an ancient glacial lake. The lake was originally formed when an arm of the glacier plugged the Wisconsin River at its previous outlet through the Baraboo Hills. In the course of centuries the lake gradually drained, choked with vegetation, and became a series of sphagnum bogs. These finally filled up with peat, which, in part, grew up to tamarack. Interspersed with the tamaracks were open bog-meadows, representing the deeper peats, and numerous sandy oak-covered ridges, representing bumps in the lake-bottom. Such was the marsh as the first white man found it.

Contrary to popular belief, the virgin marsh was probably not very productive of wildlife. Prairie chickens had not yet entered Wisconsin. There were doubtless a few sharptailed grouse, deer, and partridge on the oak "islands," but the solid tamarack stands were probably devoid of game. On the treeless bog-meadows, however, sandhill cranes and waterfowl nested in considerable numbers.

The optimum conditions for game came after settlers had begun to farm the surrounding hill country. The settlers burned large openings in the tamaracks and used them as haymeadows. Every farmer who owned a quarter-section in the hills also owned a forty in the marsh, where he repaired every August to cut his hay. In winter, when frost had hardened the marsh, he hauled the hay to his farmstead.

The open haymeadows, separated by stringers of grass, oak, and popple, and by occasional remnants of tamarack, were better crane, duck, and sharptail range than the primeval bogs. The grain and weeds on the farms abutting the marsh acted as feeding stations for prairie chickens, which soon became so abundant as to take a considerable part of any grain left in the fields. These were the golden days of wildlife abundance. Fires burned parts of the marsh every winter, but the water table was so high that the horses had to wear "clogs" at mowing times, hence no fire ever "bit" deep enough to do any lasting harm.

Then in the gay nineties came the drainage dredge, the land shark, the urge for more acreage, and all the other paraphernalia of what we fondly called "the march of empire." Ditches ten feet deep were cut through the peat, lowering the water table many feet. Farming, except in the northeast corner of the area, failed. Drain-

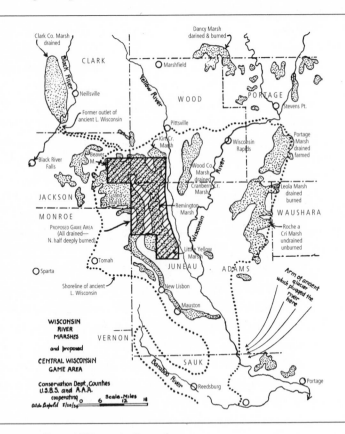

FIGURE I: Central Wisconsin Game Area

age bonds sank about as low as the water. There was, however, no immediate serious damage to the game. Drainage as such, of course, eliminated the ducks, but it sweetened the soil, augmented the growth of weeds, and thereby benefitted the chickens—until the first severe drouth allowed the dry peat to burn. From then on, during each dry period, fires literally consumed the peat beds. In 1930 a single fire lifted two or three feet of peat from an area of several townships.

These deep-burning fires present a new and puzzling problem. For the first two years after a fire the ashes produce an exuberant crop of smartweed and ragweed, which in turn produces an exuberant crop of prairie chickens and sharptails. The weedy interlude following the 1930 fires happened to fall on the high of the cycle. It is doubtful if these marshes ever carried such a stand of birds as was found there in 1932. If they did, no hunter of the present generation remembers it. Ruffed grouse also swarmed along the unburned stream bottoms. Apparently the golden age of abundant game had come back.

But not for long. The ashes were soon leached out or blown away, whereupon the food-bearing weeds disappeared, and in their place sprang up uncountable millions of aspen seedlings. Aspen is apparently the only plant capable of growing in the raw peat after the ashes are gone. By now these aspens which followed the 1930 fire are tall enough to exclude both chickens and sharptails. The raw peat on which they grow does not carry enough undergrowth to make a partridge range. It is doubtful if deer can thrive on aspen without tamaracks to winter in, and the tamarack remnants were of course largely wiped out by the post-drainage fires. So here we are with nearly a hundred square miles of aspen thicket which threatens to become, for at least a generation, a wildlife desert. To burn out the aspen would only aggravate the trouble—in fact, it would consume the remainder of the peat, expose the underlying sand, and thus create a sand dune, which might engulf the surrounding counties.

Not all of the marsh area, of course, is in such dire straits. There is much good range left, but until the fires are stopped, the entire drained area is threatened with ultimate destruction. What to do?

It is clear that the first move is to plug the ditches so as to raise the water table, and thus prevent any more deep-burning fires. But this cannot be done without money, men, and the consent of the organized landowners or drainage districts.

Marshland Economics

Most of the drained lands have passed through several salesman-farmer-sheriff cycles of ownership, and now repose in the lap of the county. Some are still held by actual farmers, some by cranberry growers, some by absentee investors. The remaining farmers are so scattered that it is excessively costly for the counties to supply them with roads and schools. It would be to everybody's advantage if these farmers holding down poor land could move to some really good land in some compact community, where road and school service would be less expensive. It would be actual economy for the public to offer them a good improved farm in exchange for the one evacuated. The AAA now offers the machinery for accomplishing just such evacuation and resettlement, if the farmers wish to use it. The CCC and FERA camps offer the labor necessary to dam the ditches, reflood the marsh, and install the food patches, plantings, and other improvements needed to check fire and restore wildlife on the evacuated lands. The Biological Survey offers to buy out the odd parcels of unoccupied holdings, but all federal participation is contingent on the state Conservation Department agreeing, once the area is completed, to operate it at its own expense. This will call for a tri-lateral "treaty" whereby the federal government and county pool their holdings, for administration by the state, as a great public "Conservation District."

Overtures looking to the consummation of this great reorganization of land-use on 100,000 acres of marsh are now under way.

The federal government and the Conservation Department are already committed. It remains to be seen whether the counties and the farmers are willing to perform their respective parts, and under what terms. It is a less expensive, less paternalistic, and more flexible plan for creating public game areas than the out-and-out federal purchase employed in such projects as the Upper Mississippi Refuge. But being a cooperative enterprise, it also requires the participation of many diverse groups, and this may not be forthcoming unless local conservation groups interest themselves in the enterprise.

Many details are still obscure, even in the minds of those in charge. It may, for example, be unnecessary to evacuate all going farms; some scattered farms may be needed to put in food patches, patrol the area, etc. Again, if the state assumes all the operating expense, it may have to seek some income from wildlife crops, especially fur crops, not so much to recoup itself as to give the county some income for the use of county lands. Will the public be willing to pay some nominal fee for the use of a public area? If not, there can hardly be any expansion of the idea in the shape of additional districts.

The local people will of course benefit, not only from the direct revenues from wildlife crops, but from the tourist trade which the existence of such crops will create. The precarious national situation in waterfowl will make it inadvisable to allow any duck shooting for a long time to come, but the upland game and fur ought to return a legitimate harvest soon after actual management begins.

Management Questions

It sounds simple to reflood a marsh and let the wildlife take it, but it isn't. Here are some of the unanswered questions:

How shall the dam-building job be divided between the engineers and the beavers? Obviously where there are no aspens the engineers must do it, also on all lower reaches where gates are needed to regulate the level of the impounded water. Can all aspen-lined headwater ditches be left to the beavers?

Will cattails tend to choke the pondage above each dam and spoil it for duck-nesting? If so, are muskrats a remedy? If so, what precautions are needed to prevent these muskrats from puncturing the dams?

With higher water-levels, will grass again get a foothold on the burned area now lost to aspen? Could snowshoe rabbits be reintroduced to thin the aspen thickets?

What about haymeadows? Isn't a partly hayed country more favorable to prairie chickens than one entirely uncut?

Do the deer need tamaracks for winter cover? If so, can such wintering thickets be artificially planted?

What grain can be used for food patches in "frostholes" where buckwheat is liable to summer killing?

White birch is necessary for winter budding of grouse, but in the heavily burned marshes is all gone. Will it grow in burned peat or sand, even if planted?

What can be done to build up the remnant of breeding sandhill cranes? No one knows what they eat, or what is the weak spot in their present environment.

Certain grasses, notably "rip-gut," are now known to form ideal chicken roosts. Where such roosts are lacking, how can the vegetation be manipulated to build up this particular grass for chicken-roost purposes?

With higher water-levels, will the acidity of the soil again increase? What changes in vegetation and animal life will this induce?

It is easy to foresee that the administration of this great area will be a technical task of no mean magnitude, requiring not only executive capacity, but exceptional skill in diagnosing these and a hundred other ecological questions. The soils, climate, plants, and animals are all unusual, so that rules-of-thumb will oftener than not prove wrong. But I can think of few tasks more worthwhile. It means restoring 100,000 acres, now destined to become a sand dune, to the varied productivity which it enjoyed in the old hay-meadow days.

Coon Valley: An Adventure in Cooperative Conservation

Leopold's criticism of fragmented, band-aid approaches to conservation shows up clearly in this 1935 article from American Forests, *which describes a novel, watershed-based effort to control severe erosion. Leopold calls here for a thoroughly "reorganized system of land-use," one that takes into account not just "soil conservation and agriculture" but also "forestry, game, fish, fur, flood control, scenery, songbirds," and all other pertinent interests. H. H. Bennett, then chief of the Soil Erosion Service (now the Natural Resources Conservation Service), was so pleased with the article that he ordered 500 copies for distribution within the agency. "You have certainly packed into this brief article a great deal of profound thought," Bennett wrote to Leopold, "and you have expressed these thoughts in a way that will appeal to the people."*

THERE ARE two ways to apply conservation to land. One is to superimpose some particular practice upon the pre-existing system of land-use, without regard to how it fits or what it does to or for other interests involved.

The other is to reorganize and gear up the farming, forestry, game cropping, erosion control, scenery, or whatever values may be involved so that they collectively comprise a harmonious balanced system of land-use.

Each of our conservation factions has heretofore been so glad to get any action at all on its own special interest that it has been anything but solicitous about what happened to the others. This kind of progress is probably better than none, but it savors too much of the planless exploitation it is intended to supersede.

Lack of mutual cooperation among conservation groups is reflected in laws and appropriations. Whoever gets there first writes the legislative ticket to his own particular destination. We have somehow forgotten that all this unorganized avalanche of laws and dollars must be put in order before it can permanently benefit the land, and that this onerous job, which is evidently too difficult for legislators and propagandists, is being wished upon the farmer and upon the administrator of public properties. The farmer is still trying to make out what it is that the many-voiced public wants him to do. The administrator, who is seldom trained in more than one of the dozen special fields of skill comprising conservation, is growing gray trying to shoulder his new and incredibly varied burdens. The stage, in short, is all set for somebody to show that each of the various public interests in land is better off when all cooperate than when all compete with each other. This principle of integration of land-uses has been already carried out to some extent on public properties like the national forests. But only a fraction of the land, and the poorest fraction at that, is or can ever become public property. The crux of the land problem is to show that integrated use is possible on private farms, and that such integration is mutually advantageous to both the owner and the public.

Such was the intellectual scenery when in 1933 there appeared upon the stage of public affairs a new federal bureau, the United States Soil Erosion Service. Erosion control is one of those new professions whose personnel has been recruited by the fortuitous interplay of events. Previous to 1933 its work had been to define and propagate an idea, rather than to execute a task. Public responsibility had never laid its crushing weight on their collective shoulders. Hence the sudden creation of a bureau, with large sums of easy money at its disposal, presented the probability that some one group would prescribe its particular control technique as the panacea for all the ills of the soil. There was, for example, a group that would save land by building concrete check dams in gullies, another by terracing fields, another by planting alfalfa or clover, another by planting slopes in alternating strips following the contour, another by curbing cows and sheep, another by planting trees.

It is to the lasting credit of the new bureau that it immediately decided to use not one, but all, of these remedial methods. It also perceived from the outset that sound soil conservation implied not merely erosion control, but also the integration of all land crops. Hence, after selecting certain demonstration areas on which to concentrate its work, it offered to each farmer on each area the cooperation of the government in installing on his farm a reorganized system of land-use, in which not only soil conservation and agriculture, but also forestry, game, fish, fur, flood control, scenery, songbirds, or any other pertinent interest were to be duly integrated. It will probably take another decade before the public appreciates either the novelty of such an attitude by a bureau, or the courage needed to undertake so complex and difficult a task.

The first demonstration area to get under way was the Coon Valley watershed, near La Crosse, in west central Wisconsin. This paper attempts a thumbnail sketch of what is being done on the Coon Valley Erosion Project. Coon Valley is one of the innumerable little units of the Mississippi Valley which collectively fill the national dinner pail. Its particular contribution is butterfat, tobacco, and scenery.

When the cows which make the butter were first turned out upon the hills which comprise the scenery, everything was all right because there were more hills than cows, and because the soil still retained the humus which the wilderness vegetation through the centuries had built up. The trout streams ran clear, deep, narrow, and full. They seldom overflowed. This is proven by the fact that the first settlers stacked their hay on the creek banks, a procedure now quite unthinkable. The deep loam of even the steepest fields and pastures showed never a gully, being able to take on any rain as it came, and turn it either upward into crops, or downward into perennial springs. It was a land to please everyone, be he an empire-builder or a poet.

But pastoral poems had no place in the competitive industrialization of pre-war America, least of all in Coon Valley with its thrifty and ambitious Norse farmers. More cows, more silos to feed them, then machines to milk them, and then more pasture to graze them—this is the epic cycle which tells in one sentence the history of the modern Wisconsin dairy farm. More pasture was obtainable only on the steep upper slopes, which were timber to begin with, and should have remained so. But pasture they now are, and gone is the humus of the old prairie which until recently enabled the upland ridges to take on the rains as they came.

Result: Every rain pours off the ridges as from a roof. The ravines of the grazed slopes are the gutters. In their pastured condition they cannot resist the abrasion of the silt-laden torrents. Great gashing gullies are torn out of the hillside. Each gully dumps its load of hillside rocks upon the fields of the creek bottom, and its muddy waters into the already swollen streams. Coon Valley, in short, is one of the thousand farm communities which, through the abuse of its originally rich soil, has not only filled the national dinner pail, but has created the Mississippi flood problem, the navigation problem, the overproduction problem, and the problem of its own future continuity.

The Coon Valley Erosion Project is an attempt to combat these national evils at their source. The "nine-foot channel" and endless building of dykes, levees, dams, and harbors on the lower river are attempts to put a halter on the same bull after he has gone wild.

The Soil Erosion Service says to each individual farmer in Coon Valley: "The government wants to prove that your farm can be brought back. We will furnish you free labor, wire, seed, lime, and planting stock, if you will help us reorganize your cropping system. You are to give the new system a five-year trial." A total of 315 farmers, or nearly half of all the farms in the watershed, have already formally accepted the offer. Hence we now see foregathered at Coon Valley a staff of technicians to figure out what should be done: a CCC camp to perform labor; a nursery, a seed warehouse, a lime quarry, and other needed equipments; a series of contracts with farmers, which, collectively, comprise a "regional plan" for the stabilization of the watershed and of the agricultural community which it supports.

The plan, in a nutshell, proposes to remove all cows and crops from steep slopes, and to use these slopes for timber and wildlife only. More intensive cultivation of the flat lands is to make up for the retirement of the eroding hillsides. Gently sloping fields are to be terraced or stripcropped. These changes, plus contour farming, good crop rotations, and the repair of eroding gullies and stream banks, constitute the technique of soil restoration.

The steep slopes now to be used for timber and game have heretofore been largely in pasture. The first visible evidence of the new order on a Coon Valley farm is a CCC crew stringing a new fence along the contour which marks the beginning of 40 per cent gradients. This new fence commonly cuts off the upper half of the pasture. Part of this upper half still bears timber; the rest is open sod. The timbered part has been grazed clear of undergrowth, but with protection this will come back to brush and young timber and make range for ruffed grouse. The open part is being planted, largely to conifers—white pine, Norway pine, and Norway spruce for north slopes, Scotch pine for south slopes. The dry south slopes present a special problem. In pre-settlement days they carried hazel, sumac, and bluestem rather than timber, the grass furnishing the medium for quick hot fires. Will these hot dry soils, even under protection, allow the planted Scotch pine to thrive? I doubt it. Only the north slopes and coves will develop commercial timber, but all the fenced land can at least be counted upon to produce game and soil-cover [*sic*] and cordwood.

Creek banks and gullies, as well as steep slopes, are being fenced and planted. Despite their much smaller aggregate area, these bank plantings will probably add more to the game carrying capacity of the average farm than will the larger solid blocks of plantings on slopes. This prediction is based on their superior dispersion, their higher proportion of deciduous species, and their richer soils.

The bank plantings have showed up a curious hiatus in our silvicultural knowledge. We have learned so much about the growth of the noble conifers that we employ higher mathematics to express the profundity of our information, but at Coon Valley there have arisen, unanswered, such sobering elementary questions as this: What species of willow grow from cuttings? When and how are cuttings made, stored, and planted? Under what conditions will sprouting willow logs take root? What shrubs combine thorns, shade tolerance, grazing resistance, capacity to grow from cuttings, and the production of fruits edible by wild life? What are the comparative soil-binding properties of various shrub and tree roots? What shrubs and trees allow an understory of grass to grow, thus affording both shallow and deep rootage? How do native shrubs or grasses compare with cultivated grasses for root-binding terrace outlets? What silvicultural treatment favors an ironwood understory to furnish buds for grouse? Can white birch for budding be planted on south slopes? Under what conditions do oak sprouts retain leaves for winter game cover?

Forestry and fencing are not the alpha and omega of Coon Valley technique. In odd spots of good land near each of the new game coverts, the observer will see a newly enclosed spot of a half-acre each. Each of these little enclosures is thickly planted to sorghum, kafir, millet, proso, sunflower. These are the food patches to forestall winter starvation in wild life. The seed and fence were furnished by the government, the cultivation and care by the farmer. There were 337 such patches grown in 1934—the largest food-patch system in the United States, save only that found on the Georgia Quail Preserves. There is already friendly rivalry among many farmers as to who has the best food patch, or the most birds using it. This feeding system is, I think, accountable for the fact that the population of quail in 1934–1935 was double that of 1933–1934, and the pheasant population was quadrupled. Such a feeding system, extended over all the farms of Wisconsin, would, I think, double the crop of farm game in a single year.

This whole effort to rebuild and stabilize a countryside is not without its disappointments and mistakes. A December blizzard flattened out most of the food patches and forced recourse to hopper feeders. The willow cuttings planted on stream banks proved to be the wrong species and refused to grow. Some farmers, by wrong plowing, mutilated the new terrace just built in their fields. The 1934 drouth killed a large part of the plantings of forest and game cover.

What matter, though, these temporary growing pains when one can cast his eyes upon the hills and see hard-boiled farmers who have spent their lives destroying land now carrying water by hand to their new plantations? American lumbermen may have become so steeped in economic determinism as actually to lack the personal desire to grow trees, but not Coon Valley farmers! Their solicitude for the little evergreens is sometimes almost touching. It is interesting to note, however, that no such pride or tenderness is evoked by their new plantings of native hardwoods. What explains this difference in attitude? Does it arise from a latent sentiment for the conifers of the Scandinavian homeland? Or does it merely reflect that universal urge to capture and domesticate the exotic which found its first American expression in the romance of Pocahontas, and its last in the Americanization of the ringnecked pheasant?

Most large undertakings display, even on casual inspection, certain policies or practices which are diagnostic of the mental attitude behind the whole venture. From these one can often draw deeper inferences than from whole volumes of statistics. A diagnostic policy of the Coon Valley staff is its steadfast refusal to straighten streams. To those who know the speech of hills and rivers, straightening a stream is like shipping vagrants—a very successful method of passing trouble from one place to the next. It solves nothing in any collective sense.

Not all the sights of Coon Valley are to be seen by day. No less distinctive is the nightly "bull session" of the technical staff. One may hear a forester expounding to an engineer the basic theory of how organic matter in the soil decreases the per cent of run-off; an economist holds forth on tax rebates as a means to get farmers to install their own erosion control. Underneath the facetious conversation one detects a vein of thought—an attitude toward the common enterprise—which is strangely reminiscent of the early days of the Forest Service. Then, too, a staff of technicians, all under thirty, was faced by a common task so large and so long as to stir the imagination of all but dullards. I suspect that the Soil Erosion Service, perhaps unwittingly, has recreated a spiritual entity which many older conservationists have thought long since dead.

Farm Game Management in Silesia*

In this article, published in 1936 in American Wildlife, *Leopold describes how farmers in the region of Silesia, then largely in southeastern Germany, managed their lands to promote game along with other crops. Although Leopold disliked some German practices of the time, particularly the massive annual practice of herding and killing an entire*

*This paper is based on studies made in Germany and Czecho-Slovakia, August to November, 1935, under the auspices of the Oberlander Trust and Carl Schurz Memorial Foundation. For help in gathering the data I am especially indebted to Gaujagermeister Freiherr von Reibnetz, Kreisjagermeister Quaschning, and Landwirt Alfred L. Schottlaender, of Breslau; Forstmeister Wilhelm Blume of Heinrichau, Graf von Rotkirch and Pathen of Massel, Graf Carl Dietrich von Haldenberg of Rettgau, Dr. Sperg of Puschkowa, Silesia; also Prince Carl von Hohenlohe-Lagenburg of Domane Rothenhaus, Gorkau, Czecho-Slovakia.

year's harvestable game crop in one "hunt," he was delighted by the widespread efforts of landowners to provide farm game with food and cover. He was also pleased by the practice of breaking up large cultivated fields with "remises": small coverts, an acre or two in size, of mixed trees planted just for game. German successes, he emphasizes here, were due not to "sportsmen passing laws and resolutions" but to landowners promoting wildlife "as a natural and widespread adjunct to good agriculture." The article carried the subtitle "How Farm Practice and Game Management Are Dovetailed."

Since World War II, Poland has controlled most of Silesia, including the region around Breslau (Wrocław in Polish) that Leopold describes here.

The American outdoorsman, hearing tall tales of abundant game in Germany, is likely to form a mental picture of a whole country hopping with furred and feathered wildlife. Such is by no means the case. In Germany, as in America, one finds farming regions which are gameless, and others which have abundant game of many species.

Why? The difference lies mainly, I think, in the local people. Some localities are willing to work for their sport, and have plenty. Others are willing merely to take what comes easy, and have little or none. *In no case does government raise game for the shiftless community.* Game officials have learned, through long experience, that the only thing they can do to improve shooting is to help those who help themselves.

While abundance thus depends primarily on local effort, there is a general trend for effort to reap greater rewards as one moves inland. This is because the soils grow better inland. The same is true with us.

Unlike America, abundance does *not* depend on shiftless farming. There is no shiftless farming. The most abundant farm game often occurs on the richest, most intensively cultivated farms. (Illinois and Iowa take notice.)

While there are gameless farm regions, there are no gameless

forests. The sportsmanship of German foresters is traditional—
they managed game long before they managed timber, hence game
management is universal in both public and private forests. At the
present time, however, forest game is in a bad way. The carrying
capacity of the range has declined due to planting too many pure-
conifer stands, which afford an excess of cover but no food.

A heavy population of deer is carried by means of artificial feeding, but without natural foods the deer herd has declined in quality and vigor. The German foresters now wish to restore a natural
mixture of hardwoods, but the deer won't let them—hardwoods
must be fenced to survive the hungry animals. The deer pressure
also tends to destroy the berries and mast which are the food supply of grouse.

There is much for American conservationists to ponder in this
situation, the details of which have been described in another
paper.*

The reason for gameless farms is the same as with us: intensive
agriculture, where conducted without regard for wildlife, has
destroyed the cover and food. A few hares and Hungarian partridges may persist on such unmanaged areas. They are leased and
hunted for what little shooting they afford, but as with us, the crop
is thin and uncertain, being determined wholly by accidental combinations of weather, gunpowder, and agricultural practice.

As with us, the shiftless sportsman on unmanaged range blames
the game shortage on "vermin," which, we are soberly informed,
hunt year-around. (I wonder when this profound deduction was
first made, and in what century it will cease to be regarded as news.)

The German farm country productive of game is remarkable for
three things:

    1. the detail with which *farm practice and game management
       are dovetailed,*

*See "Deer and Dauerwald in Germany," *Journal of Forestry* 34, Nos. 4–5 (1936),
pp. 367–375, 461–466.

2. the *variety of species co-existing on the same area,*

3. the *low per cent of artificial propagation.*

Managed farms yield a large and dependable game crop consisting of various combinations of pheasant, Hungarian partridge, hare, rabbit, and roe deer. Pheasants predominate on the heavier wetter soils; partridges in the lighter warmer loesses and sands. A heavy crop of both pheasant and partridge is seldom found on a single property. Combinations of pheasant, hare, and roe prevail on rich soils. On light soils the usual combination is partridge and hare.

One of the regions most productive of both agricultural crops and game is Lower Silesia, which lies on the upper reaches of the River Oder, and comprises that eastward projection of Germany abutting on southern Poland and eastern Czecho-Slovakia. The rich plains around the capital city, Breslau, are "the best soil in Germany," and correspond to our cornbelt. Sugar beets are the cash crop. The fields are traversed by a system of small portable tramways to haul the beets and the heavy machinery used for their cultivation. Potatoes, small grains, clover, and alfalfa comprise the remainder of the rotation. There is some dairying but no pasture, the wetter meadows being worked for hay, rather than grazed directly by stock.

On this rich central plain there is no timber except the *remises* planted for game cover, but as one climbs into the sandier rolling uplands timber progressively increases until finally, on the mountains, the country becomes solid forest.

The Silesian land-holdings are both large and small, but even where small the owners are congregated in villages. Hence the landscape in either case is one of wide horizons, broken only here and there by densely clustered habitations. There are no fences and hence no fencerows. The frequent drainage ditches are lined with a very narrow fringe of alders and willows. All highways are lined with small fruit trees. In neither case, however, is there any tall grass cover, the ditch banks and roadsides being kept short by frequent hand-mowings for hay.

The climate of the Silesian plain is similar to that of Ohio; the ground is usually bare most of the winter, but in exceptional years several feet of snow may lie for months.

Where, in such a highly cultivated landscape, shall one find shelter for game? The answer is the remise—a small concentrated spot of cover planted especially for pheasants, but used also by hares and roes. On an October evening I have frequently seen, on the fields adjacent to a single remise, as many as seventy-five pheasants, half a dozen roe deer, and a dozen hares.

A typical remise covers an acre or two, and consists of an outer belt of hedged Norway spruce, next a belt of taller unhedged spruce, then a belt of alder, and finally a central core of hardwoods, or—if on wet land—willows and cane *(phragmites)*. Fig. 1 shows the design and operation of a typical remise.

Remises are located on haymeadows or on waste spots if there

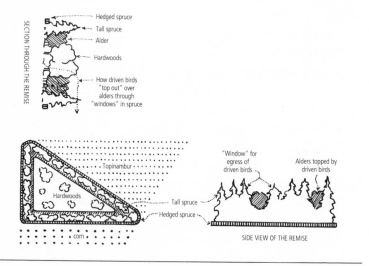

FIGURE 1: Design and operation of a remise

are any: potholes, gravel and sand pits, old slough bottoms, banks, and the like. Failing such spots, they are planted on good beet land. The Silesian farmer uses his land more intensively than we do, but he does not begrudge a few acres of good soil for wildlife remises, or for a woodland "park" around his farmstead.

The best remise system I examined had ten cover units on an estate of 780 acres, or one per 80 acres. On this estate the area in remises, park, and food patches was 8 per cent of the total. The average size of a remise was 2 acres. The average distance between remises was a long pheasant-flight, i.e., about one-third mile. Fig. 2 shows a typical estate in relation to its remise system.

When the spruce in a remise gets too large for good cover, either a new outside belt is planted, or the big trees are chopped

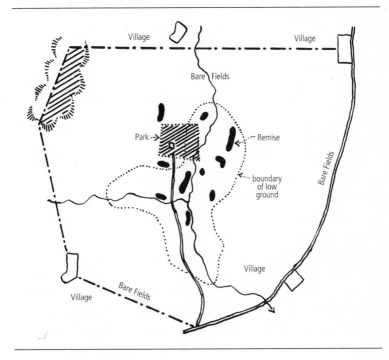

FIGURE 2: Village map

down and replaced by young ones. Such replacement is commonly done by stages, so as not to sacrifice the utility of the remise during the period when the new trees are small. Such a thing as grazing livestock in a remise is unheard of.

Some remises are laid out in unpleasing geometrical squares and triangles, but usually enough irregularity is maintained in shape, size, and content so that the system of remises greatly enhances the beauty of the farm landscape.

The remise is usually designed to facilitate drives as well as to shelter the birds, of which more later.

Most remises have a food patch adjacent or nearby. These patches consist of standing corn, and a perennial sunflower called *Topinambur tuberosa* (Jerusalem artichoke). The stem of this plant bears no seed, but the roots bear a whitish potato-like tuber, about the size of a hen's egg, which, after the corn is exhausted, is eagerly eaten by pheasants, roes, and hares. The tubers are exposed by plowing during the fall. They are not spoiled by frost. Even after plowing there are enough tubers left in the ground to form next year's crop. A topinambur patch may last as long as ten years, after which the ground must be rotated to legumes and small grains to replenish soil fertility. Topinambur is also used in forest food patches for deer.

In certain regions this plant may be the answer to the question repeatedly asked by American game managers: What perennial can one plant for winter bird food? It cannot be recommended for use in agricultural regions, however, because in this country it escapes and becomes a weed. It should be planted only in food patches surrounded by a wide belt of heavy sod.

The food patch alone does not suffice for winter feeding. Its function is to hold birds, rather than to prevent starvation. As soon as the corn is exhausted, hand-feeding begins. In selected remises straw is spread under the dense spruces, and into this wheat or barley is thrown by hand. "Automatic" hoppers are unknown (a reflection of the lesser labor costs in Europe).

All game gets winter greens from the alfalfa and clover stubbles, and in most years, acorns from the oaks in the parks.

The outstanding peculiarity of the Silesian pheasant range, and for that matter all other German game ranges, as compared with our Middle West, is the lack of high-class food-bearing weeds. Weeds not only tend to be excluded by the intensive farming and the frequent mowing of all grass, but even where they are deliberately encouraged (as was the case on one estate), they are of inferior quality—not above the grade of pigweed *(Chenopodium)*. Some day we in America will learn to appreciate our small native prairie ragweed, which comes uninvited into our stubbles, pastures, and waste spots, and the seeds of which sustain not only all farm game birds, but a host of songbirds as well. Europe has no weed food of comparable quality, and must make up for it by a more liberal use of grain and other cultivated plants.

The Silesian remises, then, offer an artificial but not wholly unnatural winter cover. I now already hear the question which springs to the lips of the American game manager: What about nesting cover? It is self-evident that the fifty pheasants which one sees feeding on the bare alfalfa stubble or newly sprouted small grains around a one-acre remise cannot possibly nest in that remise, even after their number has been trimmed down by the November shooting.

The answer is that *there is no nesting cover*, and furthermore there is no attempt to provide any. The pheasants nest in the alfalfa, clover, and grain fields, often at a great distance from any remise. In alfalfa and clover they of course sustain a heavy loss from mowing machines, just as with us. The grain, however, offers a safe nesting place, since most of it is planted in fall, grows early in spring, and stands uncut till July, by which time all except repeat-nests have safely fledged. Consequently management seeks either to force the birds out of the alfalfa and into the grain, or to salvage the hayfield eggs for artificial rearing. There are three systems in use for accomplishing one or both of these solutions of the hayfield problem.

The first consists of systematically dogging the alfalfa, clover fields, and haymeadows just before the first eggs are laid. The purpose is to drive the nesting birds out of the hay and into the grain. Any early nests found are lifted and the eggs transferred to the artificial propagation plant.

The second system is to dog the hayfields just before the hay is cut for the purpose of locating nests, and to transfer the eggs for artificial propagation. Such eggs are, of course, already partly incubated, and must be transferred speedily, set by set, to avoid heavy loss.

The third is to dog and mark the hayfield nests, and to leave islands of uncut hay so that the pheasant can complete her hatch. This system has, of course, been tried in America, but with pretty heavy crow losses. I am assured that crows are kept low enough in Germany that the system is reasonably safe, and that it is wisely used by small farmers who lack time to operate propagation plants.

Flushing bars are unknown.

Propagation systems in Silesia are identical with ours, except that mechanical incubation and brooding are rare, and turkeys as well as chickens are used as foster-mothers. Rearing fields are commonly not placed on clover, but rather on haymeadow. Pens contain live growing spruces as cover. The turnips grown as a soiling crop in pens come right up to the stems of the hedged spruce trees.

Usually only hayfield eggs are artificially propagated, but a few holdings produce and propagate additional eggs, and I saw at least one which operates as a commercial game farm. Other large holdings depend wholly on natural propagation. The commercial game farm which I saw used turkeys for foster-mothers, reared in open fields with "A" coops, let the young go wild, shot only cocks, and then *trapped the excess hens* for sale. This wild stock is preferred for restocking purposes.

In the Breslau district the ratio of released artificial birds to annual kill averages 1:10, i.e., the bulk of the kill is wild stock. Holdings operated by small farmers, of which there are 186 units or shoots in the Breslau district alone, never resort to artificial propagation, but many of these nevertheless have good stands of pheasants (see Unit B, Table 1).

TABLE 1: Kill on single estates, Silesia

Estate or Unit	Area in Acres	Kill				Art. Prop. Pheas. Release	Acres per Bird		Per cent in Remises
		Pheas.	Part.	Hare	Roe		Stand	Kill	
A	5,500	3,500[1]	?	?	?	Hayfield eggs only	0.5	1.6	8%
B[2]	900	200	500	280	?	None	?	1.3	?
C[3] (1934–35)	23,000	3,442	107	1,549	131	Many	?	6.5	Large
D (1933–34)	2,800	654	126	206	14	500	?	3.6	Large

[1] 3,000 cocks, 500 hens. The fall stand on this estate is 8,000 pheasants, 300 partridges, the kill a little under 50 per cent.

[2] A "cooperative" owned, managed, and shot by small farmers.

[3] Poor, sandy soil. Figures are for season 1934–35, a poor pheasant year. Part of kill is trapped birds sold for restocking.

Up to a few years ago the blackneck pheasant *(P. colchicus)* was standard stock in Silesia, but the present preference is for a cross between the ringneck *(P. torquatus)* and the Mongolian. This change in species is being accomplished by the periodic release of outside breeding stock for "new blood."

The present prejudice against blacknecks is grounded on the alleged wandering tendencies of this species, but I think the greater meat poundage of the larger birds also comes into play. Despite the introduction of the larger species, the average pheasant one sees in the fields seems to me distinctly smaller than ours, i.e., to run heavier to blackneck blood.

To an American one of the most surprising and least pleasing aspects of Silesian pheasant management is that most of the pheasant shooting on each holding is concentrated into a single big one-day drive, during which very large bags are made. If any small hunts are made, they are made sparingly by the owner, and usually after the big shoot. This custom of concentrated shooting rests on experience: It has been found that to make frequent small hunts pushes the birds over the boundary to the neighbors, so that the landowner fails to realize on his own crop. Our states which have enacted shooting preserve laws are, of course, experiencing this identical difficulty.

This tendency to embark for parts unknown at the crack of a gun constitutes, I think, the basic weakness of the pheasant as a game bird for America: He tends to be inherently unsuited to our preference for small-but-often dog-hunts. In Silesia where there is no standing corn, no grass, and no marsh to help hold the birds, the scattering tendency is probably even worse than in the unman-aged ranges of Ohio, Iowa, or Wisconsin.

Sometimes a landowner, in order to drive his birds inward, does a little advance dog-hunting on his outlying boundary-line fields which have no cover, but he is always careful to burn no powder near the remises until the big day of the annual shoot. On that day the birds are so herded from remise to remise by skillful beaters, that to the pheasant one direction probably seems no better than another for get-away purposes. I imagine that, to a pheasant, a Silesian drive is on all fours with Opening Day in an Iowa corn-field.

This brings us now to an important point. The "big drive" con-sists in pushing the birds from one remise to another, over the heads of the intervening line of guns. The remise system, however, may cover only a fraction of the hunting unit, and lies usually *at its centre*. That is to say, the hatch of birds from the whole unit is, by virtue of the distribution of remises, *crowded into a small winter range*, on which all drives, all winter cover, and all winter feeding are concentrated. This winter range is always the lowest, wettest, flattest ground. It follows that the outlying borders of a hunting unit are usually bare fields habitable during summer only, and that this bare zone tends to act as a barrier to discourage scattering dur-ing the fall hunts. Fig. 2 illustrates such a central system of remises.

Agricultural practice is also modified to push birds inward and to discourage dispersion. Thus of any two fields ready to cut at one time, the outer is always cut first.

To sum up, the birds are pushed inward by peripheral shooting and cutting, by centripetal feeding and cover, and, during the big shoot, by centripetal driving.

Table 2 shows the average kill in the Breslau district, which is as large as a small county. The kill is 1 bird (pheasant or partridge) to each seven acres. The average kill on the intensively managed estates shown in Table 1 is 1 bird to each two acres for rich land, and 1 per five acres for sandy land. I am assured that just about half the fall population is shot, so one may say the fall density on the best estates is a bird per acre, counting both the summer and winter range. The density in the winter range or remise area is of course much greater; I saw 448 pheasants shot in one day out of eleven remises of not over three acres each, a kill of 15 birds per acre of remise cover. Silesia, in other words, achieves a stand of pheasants equal to South Dakota and better than north Iowa, without the standing corn and long grass which make a heavy stand in those lucky states nearly automatic. One may say further that this heavy stand is 90 per cent natural, and grows on machine-farmed land in the face of a human population-pressure heavier than we know anything about. This, I submit, is a substantial achievement.

TABLE 2: 1934 kill in Kreiss Breslau 230,000 acres

Game	Kill	Acres per Head	Predators and Fur	Kill	Ratios
Roes	1,086	} 11.3	Fox	49	
Hares	19,114		Badger	3	
Pheasants	22,378	} 7.0	Marten	11	
Partridges	13,602		Weasel	832	
Woodcock	26			895	Game per mammal 64:1
Jacksnipe	15				
Pigeons	207		Accipitrine hawks	176	
Ducks	1,055		Buteo hawks	69	
Total	57,483	4.4		245	Game birds per raptor 142:1
			Crows and magpies	2,021	
			Total	3,161	Game per predator 18:1

I cannot help adding that this achievement must be credited not to sportsmen passing laws and resolutions for each other's governance, but to landowners who go out on their land and practice game management as a natural and widespread adjunct to good agriculture.

The Silesian landowners and officials seem unanimous in the opinion that a breeding ratio of 1 cock:5 hens or 1 cock:6 hens is right for wild pheasants. They believe in this so explicitly that when more cocks than this are left over they often—with the permission of the state warden or *jagermeister*—go out in spring to shoot off cocks which seem to be without harem. This belief seems to run counter to the earlier conclusions reached by Wight, who found in Michigan that harems of over two to three hens per cock are rare. His later work, however, seems to leave the maximum hens per cock as an open question, and one probably without any fixed and uniform answer.

During the big drive, it is the usual practice to shoot only cocks. In certain remises, however, the owner may pass out word to shoot hens; this is because a previous census has shown that particular remise to have more hens than are needed. Later in the season the owner, his game manager, or a few invited officials—all experienced men—may hold one or two small shoots in which old hens are trimmed out.

The total kill is about half of the fall population. The resulting sex-ratio after the hunting season may be 1:6, but the winter losses falling heavier on hens will make the spring ratio 1:5. As already stated, new stock is often released during the winter for "new blood," but this is done in the desired ratio, and does not change the sex composition of the population. Wild trapped stock, rather than game-farm stock, is usually used for new blood. It is acquired sometimes by purchase, sometimes by exchange of trapped birds between estates.

On the big estates managed intensively for pheasants there are

few partridges. On the sandier lands, however, partridges are sometimes more numerous than pheasants, especially where there are not
remises and hence no pheasant cover. For the Breslau district, the
partridge kill is about half as large as the pheasant kill (see Table 2).

The partridge crop fluctuates much more violently as between
years than the pheasant crop. This fluctuation will be discussed in
a separate paper on the game cycle. A German study of this fluctuation has recently been made by Nolte.*

It is my impression that the partridge crop is more of an accidental crop than the pheasant crop. The only management measure generally practiced is winter feeding. In remise country, partridge feeding is combined with pheasant feeding by placing the
straw beds already described near the edge of the remise, where
partridges will not hesitate to enter. In open country without
remises, special straw beds for partridges are placed in the open.

Partridge shooting is done mostly in late August and early September, before the pheasant broods have separated, hence the partridge shooting does not scatter the pheasants, which at that season are not much disposed to wander. Some partridge shooting is
individual dog-work, the rest of it "walking-in-line." Rarely both
pheasants and partridges are shot together in October and November line-hunts. This, however, is resorted to only by landowners
who do not hope to hold their pheasants, but on the contrary want
to shoot a few before they have all departed for their neighbor's
remises.

I saw one walking-in-line partridge-hunt in Czecho-Slovakia
in which about 1,000 birds (not counting reflushes), 180 hares, and
4 pheasants were put up on about 1,200 acres. The bag was 278 partridges. I think this stand was about a bird per acre, and represents
the best of the range on the Bohemian plain. There is no winter
cover. The birds nest in alfalfa and winter wheat. They are fed dur-

*W. Nolte, *Zur Biologie des Rephuhns,* reviewed in *Wilson Bulletin* 47 (December
1935), pp. 300–303.

ing snow, but there is no other management except predator control and regulation of the kill.

Hares may be called an accidental by-product of pheasant and partridge management, although they occur even where there are no birds. They eat the topinambur and grain set out for pheasants, and benefit by the remise cover, but their mainstay is the clover and alfalfa.

Most hares are shot in separate line-hunts or enclosing-drives, the latter resembling the jackrabbit drives of our western states. Just as with our rabbit, the late fall hare drives are regarded by German sportsmen as a sort of light comedy, which forms an agreeable contrast to the more serious job of hitting pheasants or partridges.

The American reader must grasp the unfamiliar fact that the roe deer in Silesia, while he uses remises and woods, is not dependent on them. There are "field roes" which never see the inside of the remise or any other cover for months at a time, and which wander about in the open fields like antelopes. Even the fawns may be dropped in treeless clover fields, and when thus cached in hay they get caught in the mowing machine like partridges or pheasants.

The Silesian pheasant remises, scattered over a fertile fenceless plain of beets, clover, and alfalfa, are ideal roe range. On several evenings I counted ten to forty feeding in the fields. Roes feed on the pheasant food. A special kind of leafy non-freezing cabbage is also planted for their benefit. Roes rut, fawn, and shed much earlier than other deer, hence the bucks are hunted in summer, usually by waiting at evening in a *hochsitz* or elevated blind on the edge of woods.

Our small woods in America have no really first-class game mammal like the roe, but we are even-up when we consider that small woods in Germany have no really first-class game bird like the ruffed grouse and the quail.

American wild turkeys were planted in Silesia some twenty years ago, but failed.

Table 2 indicates no excessive kill of predators: 1 hawk killed per

142 game birds killed; 1 predator per 18 head of game for Silesia as a whole. I fear, though, that in respect of raptors at least, the explanation lies in their near-eradication. I saw only 1 hawk during a week's travel in Silesia (and only 45 hawks in three months in Germany).

It is also possible that in making their reports to the game officials, the game managers may "go easy" on their predator kill. The game:predator kill ratios I gathered from private shooting books, in which most of the entries pre-dated the present *Naturschutz* movement, run between 1 and 10 head of small game killed per predator killed. In these books the local game manager is showing the owner that he is "on the job." This question is more fully discussed in a separate paper.*

The bulk of the predator kill consists of crows, of which there are two common species. Every spring the German sporting magazines bristle with advertisements of crow poisons, used mostly in eggs. Despite such ruthless control, crows continue numerous. One recent writer thinks the extermination of horned owls has given crows an undue advantage. If a hunting unit fails to poison its crows, the game officials may, under the new game law, require that this be done.

The German attitude toward foxes is in process of change. On some properties foxes have long been managed on a sustained-yield basis. A new law now prohibits the use of steel traps. The first result was an alarming rise in the fox population, and a loud hue-and-cry about loss of game. The second result was a sort of renaissance of ingenuity in killing foxes without steel traps. Every issue of the German sporting press now brings forth new legal traps, new calls, new baits, and new methods of driving or stalking foxes. Fox shooting is coming into its own as one of the really difficult high-class field sports. A parallel trend toward sporting fox shooting is clearly apparent in Ohio, Iowa, and Wisconsin.

*Aldo Leopold, "*Naturschutz* in Germany," *Bird-Lore* 38 (March–April 1936), pp. 102–111.

It must be conceded, I think, that the production of a heavy game stand on such a nearly coverless range as Lower Silesia probably requires a more radical predator control than is necessary or even desirable on well-covered moderately populated game ranges in America. This is merely one of a dozen basic points where we, by reason of our good luck in having more room, can improve on European conservation practice, *provided* our landowners can muster the persistence and enthusiasm which has made a sportsman's paradise of Lower Silesia.

Be Your Own Emperor

This delightful piece, subtitled "A Report on the Progress of Wild Life Cropping in Wisconsin," dates from around 1938 and is published here for the first time. Beneath its lighthearted narrative of hunting history, from ancient Rome to feudal Europe to the time of Theodore Roosevelt, it offers a sobering assessment of the challenges facing wildlife management on public and private lands. Leopold shows heartfelt concern for "rare or unprofitable species" and for migratory game that no owner can control and protect. And he bemoans the still-potent "pioneer tradition" of dominating nature, a tradition, he says, that "agricultural colleges have aided, abetted, and intensified." As he often did, he ends up calling for more practical research and for science-based practices tailored by landowners to the local land.

IT IS NOW five years since this university first began to discuss wild life as a possible land crop. It should be time, therefore, to take stock of the successes, failures, and trends of the enterprise to date, and its needs and possibilities for the future.

To do this, we should first understand something of the history of game cropping and how the present Wisconsin enterprise differs from that of other states, and from its historical counterparts in Europe and Asia.

History

Game cropping was first practiced in the Holy Roman Empire in the 12th century. Frederic II, who has been called the first scientist in Europe, had an elaborate game-cropping system in Sicily. He hunted entirely with hawks, and his "kind of game birds" was the crane. The descriptions of his operations are all in Latin and are widely scattered through a large literature, hence no one has dug them out. I wish I could be the graduate student who exhumes and reconstructs the Sicilian picture.

In the 13th century an equally elaborate system was practiced by the Mongol emperors in eastern Asia. Marco Polo has left us a brief but convincing account of its methods and results. The methods included food patches, feeding stations, closed seasons, wardens, and all the paraphernalia of a modern conservation department. The results were an annual hunt by the whole imperial court, plus "camel-lods of birds" for the imperial table throughout the winter season.

In the 15th century Edward of York records the beginnings of the present European technology. His "little simple book," "Master of Game," dedicated to the heir of his sovereign, Henry IV, is a delightfully naive exposition of that oldest of all mutual admiration societies—the sporting fraternity. Hearken to Edward:

> This book treateth of what is to every gentle heart the most disportful of all games, that is to say hunting—which is so noble a game, and lasts through all the year of divers beasts that grow according to the season for the gladdening of man. —Hunting causeth a man to eschew the seven deadly sins. —When a man is idle, he abides in his bed or in his chamber, a thing which draweth men to imaginations of fleshly lust and pleasure. Such

. . . men think in pride, or in avarice, or in wrath, or in sloth, or in gluttoning, or in lechery, or in envy.

(But) when the hunter riseth in the morning, he heareth the song of the small birds, the which sing so sweetly with great melody and full of love . . . and when the sun is risen, he shall see fresh dew upon the small twigs and grasses, and the sun by his virtue shall make them shine. After when he shall go to his quest—he shall meet anon with the hart without great seeking, (which) is great joy and liking to the hunter.

And when he cometh home he cometh joyfully. —He shall doff his clothes and his shoes and his hose, and he shall wash his thighs and his legs, and peradventure all his body. And in the meanwhile he shall order well his supper, with roots, and of the neck of the hart, and of other good meats, and good wine or ale. And when he hath well eaten and drunk he shall be glad . . . and at ease. —And then he shall lie in his bed in fair fresh clothes, and shall sleep well and steadfastly all the night without any evil thoughts.

—To be idle and to have no pleasure in either hounds or hawks is no good token.

I could prove from this many things irrelevant to game cropping. In fact, I could prove almost anything, from the fact that hunters originated the bath, to the fact that only woodsmen are virtuous. But this has been done annually in every address to sportsmen from the days of Xenophon, through Teddy Roosevelt, to the last national convention of the Izaak Walton League. So let us return to our knitting.

All three of the historic expansions of the art of game cropping—the Sicilian, the Mongolian, and the European—are alike in that their techniques were empirical, not scientific. Hence no matter how well they worked, the techniques could not be translated to other countries. This may answer the persistent question: "Why don't you get your game management from Europe?"

All three were also alike in that game was produced for private pleasure. The objective in each case contained not even a trace of the present idea of public recreation. The one-gallus hunter dates from the days of Robin Hood, but the notion of giving him any political recognition is very modern indeed.

All three were also alike in that the farmer had no place on game lands—he was either expropriated or overridden.

Lastly, wild life other than game did not enter the picture.

Hence the present attempt to crop game, while superficially similar, is really radically unlike what Frederic, or Kublai Khan, or feudal Europe did in centuries past. It is an attempt to graft together five historically separate and hitherto antagonistic elements: the Edwardian idea of wholesome sport, the democratic idea of public recreation, the esthetic idea of nature-study, the economic idea of diversified and mutually supporting crops, and the scientific idea of cropping techniques based on biological science. It is, in short, an attempt to encourage the American farmer to be his own emperor.

So much for the pipedream. What progress is it making?

Between the time of Robin Hood and the American Revolution, there germinated, grew, and flowered a deep-seated revulsion against special privileges, including hunting privileges. In our pioneer days of abundant game, this idea threw out its roots like a green bay tree. When the game played out, the American hunter concluded that hunting must vanish from civilized countries. Even up to the time of Roosevelt it clearly was assumed that the game must eventually go. But pretty soon he had a new idea, wholly original and wholly indigenous to this dark and bloody ground we live in. Government, he found, could do anything. Why not pass a law to let government perpetuate his game supply? We have spent fifty years finding out that game cropping is one of the things government cannot do, except on its own lands.

The idea of free public game dies hard. The average conservationist still believes that government can, by some waving of legislative wands, spontaneously generate a game crop, to be harvested by all comers, on lands not owned by them, and on lands increasingly devoid of food and cover fit for wild life. As long as the public believes in and supports this legend of an official Santa Claus, our public officers are not likely to suddenly disclaim the role. Let us try to reappraise the fable to see what degree of truth it may contain.

Wild Lands

The salient new fact bearing on this question is the universal admission that our low-grade lands not only are wild, but will remain so, and that they are *coming into public ownership* by the automatic process of reversion for taxes. To this extent the fable of free public game is no longer a fable, but a very real opportunity. We have or are getting the land—all that remains to be done is to prepare it for game cropping and to regulate the harvest to its productive capacity. Are we ready to do this?

In Wisconsin there are two main owners of this wild domain, and we shall have to consider them separately.

The first is the county. The average county obviously is not prepared to itself engage in a technical land-cropping enterprise. It has two ways out: Lease the land to private enterprise, or entrust it to the Conservation Commission. The latter seems to be favored. There are now evolving two vehicles for state cropping of reverted county holdings: the "county forest" and the "state conservation district." Both are promising. On a trial area of some 120,000 acres in central Wisconsin, we have the somewhat unusual spectacle of active federal participation in an enterprise to be run by the state. The AAA is buying out the scattered farms, while the FERA and the CCC are building the needed dams to crop game.

The second modern land-baron is the Forest Service. Unlike the county, the Forest Service has a field force and hence can crop its own holdings. Until this year, however, it has entirely neglected its opportunity, resting upon the antiquated assumption that game automatically grows wherever silviculture is practiced. It is now making a very spirited effort to build up a game personnel and a cropping technique.

In my opinion there are five weak points which are most likely to retard progress toward the successful cropping of wild public lands:

1. *The lop-sided and inadequate research program.* Neither the state nor the Forest Service are spending one cent for laying a scientific foundation for this huge venture. The only game research now under way in Wisconsin is carried by the university, and deals only with grouse. None of the other wild-land species, such as deer, trout, or wild furs, have been studied, here or elsewhere.

2. *Conflicting techniques.* There is still a sad lack of coordination between engineers, foresters, and game men, due not to any lack of mutual good will, but simply to lack of mutual understanding. For example, after having completed a gridiron of fire lanes opening nearly every wild spot in the north woods to motor travel, it was suddenly realized that these roads threaten to deplete our deer. An effort is now being made to put locked gates on them. I hope it will work, but I doubt it. A breachy cow has nothing on an excited hunter with a car.

3. *Lack of technical personnel.* The Conservation Commission has virtually the same technical overhead as it had five years ago, to carry responsibilities easily ten times as great. These men literally have no time to think—a dangerous condition when history is being made daily.

4. *Lack of control of public use.* No one has yet devised a way to spread the public over public lands in such a way as not to ruin the land, as well as the recreation. The present intolerable concentration of deer-hunters, for example, will not be corrected by making the land a county or national forest. This problem remains a thorn in the rose of subsidized recreation.

Despite these delays and discouragements, it is a clear fact that the idea of free public game has, in respect of wild lands, taken a new lease on life.

5. *Poor interspersion.* The reverse is true of farm game.

Farm Lands

Contrary to usual belief, the farm lands of Wisconsin, by reason of their richer soil, are capable of producing more man-days of hunting recreation than the wild lands. A stand of 1 game bird per 4 acres and a kill of a bird per 12 acres is easily possible. This would yield, on 12,000,000 acres of farms, a kill of 1,000,000 birds per year. There are 200,000 licensed hunters, so our farms could produce 5 for each licensed hunter. The maximum stand of wild birds on farms would be only thrice as great, or 15 per hunter. The additional yield of wild-land birds might double this. Hence those who fear future overproduction may forget their worries. The problem is to get the crop grown. There is no bugaboo of too much.

Our farms as they now stand are operating at perhaps a tenth capacity. A tally of 600 farms made in the winter of 1932 showed only 1 out of 4 habitable for game. Half were devoid of both food and cover, a quarter had only the one or the other, and a quarter had some of both. But those which had both seldom had enough to realize the capacity of the land. Game cannot live in quarters provided only with beds, or only with dining rooms, or with neither. All animals, including ourselves, need both. When one or both are lacking, reproduction and survival decline. Population shrinks until it fits the habitable fraction of its environment. That fraction on Wisconsin farms is 10 per cent or less.

How to put food and cover on the other 90 per cent? This is the very heart of our problem. All the conservation laws and dollars and commissions and wardens and speeches in Christendom cannot put game, or any other wild life, on fields bare of food and cover in winter.

Many short-cuts to this end have been proposed. Most of them are, I think, spurious. One faction would have the state cede private title to the game, and by giving the farmer a wide-open market for both shooting privileges and meat, make game so profitable that it could compete outright with other crops. But what would this game be? Mostly pheasants. What would become of rare or unprofitable species? They would likely disappear. With open markets for resident game, what would become of migratory? I will leave you to guess.

Nevertheless there is an intrinsic merit in the idea of creating an economic incentive for game cropping. The American Game Policy proposes to encourage the sale of shooting privileges, but not of meat. The state retains title, but the farmer is its custodian, and is encouraged to crop the resource, subject to control by the state. This, I think, is sound.

But there are serious obstacles to getting started. One of the most serious is the short seasons made necessary by the progressive failure of the public Santa Claus idea. Open seasons on upland species in the north central region averaged 60 days in 1905, 45 days in 1915, 29 days in 1925, and 14 days in 1934. The producing farmer cannot market his shooting in a week or two. Neither can the state relax its restrictions on unmanaged farms for the sake of encouraging the few managed ones. The way out seems to be a differential season, long for lands on which managed game crops can be shown, short for other lands.

The Wisconsin Shooting Preserve Law establishes such a differential. It is an opening wedge, so far applicable only to artificially raised pheasants. The state says: "For every 100 pheasants you turn out, we will let you shoot 75, and give you four months instead of four days to do it." We now have forty such preserves. There would be many more, but for the high cost of chicken-wire production. It costs $1.50 to raise a chicken-wire pheasant, perhaps 15 cents to raise a wild one. The wild bird is superior in quality.

Why is the differential season limited to liberated pheasants? Because they can be counted out by the warden. Our wardens have been too busy chasing poachers to learn how to census a wild game crop—a task often tedious and calling for skill, for favorable weather, and for dogs. Iowa, though, is doing it on quail. I predict the gradual spread of the differential season to all kinds of wild resident game. What it amounts to is that the state delegates its restrictive functions to the producing landowner, subject to cancellation for abuse. This is sound principle, and tends to relieve the wardens of the perfectly impossible job of policing game singlehanded. Moreover, it opens to the farmer, or through him to his sportsman friends, a powerful incentive to *earn* additional shooting privileges by growing a crop capable of withstanding them. The old system, on the contrary, makes shooting easier to steal than to earn. It contains the seeds of its own death.

I would not, however, leave the impression that game laws or farmer-revenue are the only or even the principal means to wild life restoration. The problem is not so simple. An even more basic reason for the modern foodless, coverless, lifeless farm is the pioneer tradition that uncut or ungrazed brush, weeds, grass, or timber bespeak an incomplete victory over the wilderness, and that any vestige of them in gully, rockpile, bank, or fencerow brand the farmer as a sloven. It is an indisputable fact that the agricultural colleges have aided, abetted, and intensified this tradition, sometimes with sound scientific reason, but more often out of that same blind subservience to fashion which dictates the color of our hats, the height of our heels, or the length of our coattails.

I plead that the wild life cover, at least on waste corners and fencerows, now become an expression of localized scientific reasoning and the owner's personal taste, rather than a badge of compliance to social regimentation. I submit that the slick and clean countryside is neither more beautiful, nor—in the long run— more useful than that which retains at least some remnants of nondomesticated plant and animal life. The brushy fencerow and the wild-grown bank, the clean-boled woodlot and the undrained spot of marsh—I heartily agree that the presence of these things on the farm portray the character of the owner, but in my view they portray him not as a sloven, but as one who, despite the stampede of his neighbors, has refused to trade his birthright as a husbandman of living things for the shoddy imitation of a factory.

It must be confessed, in all fairness, that our tax laws, and incidentally our highway engineers, have contributed their quota to the creation of the American steppe. The farmer who uses tillable or grazable land for conservation does so at his own cost, but the lumberman who sacrifices the immediate personal penny to the ultimate public good gets a handsome subsidy for public services performed. Is not a marsh, a prairie, a food patch, a grape tangle, a grove of veteran oaks, a patch of ladyslippers, or even a copse of flowering haws a public service in the same sense as a commercial forest? There must be some sound way in which the public that wants these things to exist can ease the economic pressure, which now tends to wipe them out of existence. The public is spending

millions to perpetuate such things in public parks and reservations, but what kind of "landscaping" is this? These things should not be crowded into distant reservations. They belong on farms, where they can be seen by everybody every day. The esthetic dietary implied in parks is too much like that of the starving prospector who ordered forty-dollars worth of ham and eggs. Strung out a bit now, they might do more good.

Game Technology

We have in Wisconsin five species of farm game, four of wild-land game, half a dozen of fur-bearers, and a dozen of migratory birds; total about twenty-five species.

Most of these are as distinct in their cultural requirements as alfalfa is distinct from corn or oats. Each has its distinctive kinds of food and cover, in certain proportions, at various seasons. Each has its diseases and enemies. It takes a good research man four years to discover the rudiments of a technology for each. This adds up to a total liability of 100 man-years of research to get started in Wisconsin game cropping.

We have so far had about 10 man-years of research. Fortunately they have been effectively spent. We now know the rudiments about quail and prairie chickens. Michigan has done something on pheasants, Minnesota on ruffed grouse. Iowa has made a start on two ducks. We can borrow this information until we have better. But of deer, cottontail, all the fur-bearers, and most of the waterfowl, we know, biometrically speaking, nothing.

What little we do know has neatly upset many time-honored beliefs, among them the belief that one less hawk means several more game birds. I am beginning to believe that Errington's discoveries in predation, largely based on his Wisconsin quail data, are of general importance to science as well as to game management. He finds, in short, that predators trim down the quail population to the capacity of the food and cover, *but no further*, and that this shrinkage *occurs anyhow,* regardless of whether predators are abundant or scarce, or of what kinds they are. In other words, the only effective predator control is food and cover improvement. "Vermin" baiting is, apparently, a needless waste. These indications, if substantiated, are a fundamental contribution, not only to conservation policy, but to ecological science.

PART II

A Landowner's Conservation Almanac

Leopold wrote the forty short essays in this part between 1938 and 1942 and intended them all for publication in the Wisconsin Agriculturist and Farmer, *a widely read farm newspaper. Of these, seven have never before been published: "Do We Want a Woodsless Countryside," "The Farmer and the Fox," "Pines above the Snow," "Roadside Prairies," "The Hawk and Owl Question," "Smartweed Sanctuaries," and "From Little Acorns." Leopold had several aims in mind as he addressed his practical-minded, hardworking readers. He wanted to describe useful conservation practices and to encourage readers to implement them, whether in feeding quail, constructing ponds, or allowing dead trees to stand. The suggestions he offered were easy to implement and required little or no cash outlay; if enough readers implemented them, he suggested, the landscape would be transformed. Leopold also wanted to get readers to enjoy the outdoors far more, appreciating its natural beauty, taking pleasure in its many creatures, and sensing its alluring mystique. Over the long term, he hoped, readers would love the land and sink deep roots in their chosen places, just as he had done.*

Leopold knew, though, that a mere love of natural beauty would not equip readers with the tools they needed to use land wisely. They also needed to gain knowledge about the land. They needed to get to know their plant and animal neighbors and to notice when a species declined or disappeared. They needed to learn about the natural processes that formed the land and kept it fertile. Most of all, they needed to gain the kind of curiosity about nature that would keep them attentive, year in and year out, to the entire land community's ways and well-being.

Finally and most ambitiously, Leopold sought to reshape subtly his readers' values and aims. He wanted them to develop a historical perspective on the land so that they might value more highly what existed and plan more effectively for the long term. He wanted them to take pride in owning and nurturing a diverse, beautiful land. And he wanted them to give thought to why they lived in rural areas, what they valued, and what made their lives worthwhile. If landowners could ponder such matters, he believed, they would be more likely to buck economic pressures and pursue holistic aims.

Leopold's essays succeed not just because he knew his material but also because he respected his readers. He let them decide how many rabbits to produce, which birds to attract, and even whether to watch animals or shoot them. He opened his own heart and displayed his own preferences, but in the end, he knew, landowners had to make their own choices.

WINTER

Winter Cover

THE RECIPE for wintering game is corn and cover. Corn is easily provided but cover is often scarce. This tells how winter cover may be improved.

By winter cover is meant vegetation stiff and tall enough so that snow will not bury it, and thick enough to protect birds from weather and enemies.

Young evergreens, grape tangles, and bushy marsh are the three kinds of cover dependable in all weathers.

Bushy Marsh. Nothing excels a bushy marsh as winter cover, especially for pheasants and quail. After snow and ice have flattened down all grass and weeds, the bushes still stand up. Willow, red dogwood, and alder are good bushes.

All marsh tends to grow bushes if protected from fire, grazing, and mowing. Marsh cover, then, is a matter of leaving part of the marsh unburned, ungrazed, and uncut. If the marsh is drained, ditch banks are a good place to leave bushes.

To hurry the natural process of growing bushes, plant cuttings. This is best done in spring.

Uncut cattails, slough grass, or sweet clover are valuable to reinforce the bushes and to serve as roosts, but are not weatherproof in themselves.

Grape Tangles. A grape tangle can be grown in a single season, and will house more game per square rod than any other cover.

One way to build a tangle is to find a tree with a vine in its top. Fell the tree, but don't cut the vine. Leave the stump high, with the butt resting on it. The vine soon converts the down tree into a dense tangle.

Another way is to find a grape tangle sprawling on the ground. Build over it a "teepee" of stout durable poles. Wrap the teepee with discarded wire. Pick up the vines from the ground and lace them into the superstructure. The vine soon covers the whole structure with dense growth.

Grape tangles, to grow vigorously, must be in full sunlight. They must be protected from livestock.

Evergreens. These are best planted in spring. Special directions will appear in a spring issue. Brushpiles, and down tops of oaks felled with the leaves on, are good temporary covers while waiting for evergreens or grape tangles to grow. Brushpiles are more effective if topped with cornstalks or marsh hay.

It is useless to plant birds on a farm without good cover. New fences may bring more birds than new law.

Winter Food

THE RECIPE for wintering game is corn and cover. As in other recipes, success depends on where, when, and how "the makings" are put together.

Where? For quail, put your corn on the leeward (south or east) side of the thickest cover, preferably on the sunny side of a south-facing bank. Quail dislike to feed in the open.

For prairie chickens and Hungarian partridge, put corn in the open, never in woods or brush. Grass or weeds nearby are good but not essential.

For pheasants it doesn't matter where corn is put, provided there is plenty of it.

Wastage by squirrels is serious when corn is left in or near woods. A brush patch, marsh edge, or fencerow is therefore a better place than a woodlot to feed quail or pheasants.

When? Start early, preferably by November. This is especially important in order to hold quail and pheasants on uplands. If no corn is offered until the winter storms actually start, your birds are likely to drift to some distant marsh and you may never see them again.

How? The best and simplest way to feed is to leave some shocks or standing corn in the field. If shocks are used, new bundles should be exposed from time to time in winter. Don't worry about birds learning to remove husks. Even quail become expert at this. The next best way is to put corn ears in a wire basket or impale them on spikes in a pole (see drawing).

Shelled corn is best fed in a hopper, but the hopper must be roofed over, else rain or melting snow will clog it with ice. Where squirrels are present, the tray of the hopper should be covered with wire mesh to reduce wastage.

The manure-spreader is a good feeder, especially for pheasants and Hungarians, but the danger is that spreading will cease in bad weather when the birds need it most.

The poorest way to feed is to force the birds to visit the barnyard. Dogs and cats harass the birds. In quail this may be fatal, especially if the covey fails to reassemble before night.

How Much? In a hard winter a quail eats half a pound of corn per week, a Hungarian three-quarters of a pound, a pheasant or prairie chicken two pounds, a rabbit one pound, a squirrel two pounds. This is over and above the wild foods which they "rustle" for themselves.

It takes a tenth of an acre of corn to support a good stand of game on a Wisconsin farm.

Substitutes for Corn. Soy-beans, small grains, or foxtail and ragweed seeds gathered under the shredder are all good winter food. Weed seeds must be dried, else they will mold. Small grains are best fed in shocks or stacks.

Winter feeding is food for people as well as birds. It is no small satisfaction, when blizzards howl, to know that your birds have fuel in their stoves.

Feed the Songbirds

Would you like to sit at your south window and watch cardinals, chickadees, nuthatches, juncos, tree sparrows, bluejays, and woodpeckers eat breakfast?

It is easily arranged.

First, keep a piece of suet tacked on some nearby tree. This is for the chickadees, nuthatches, and woodpeckers. If the jays carry away too much of it, cover it with wire mesh.

Second, erect a feeding tray; any flat surface set on a stump or post so as to discourage cats. Keep this clear of snow and sprinkled with cracked corn, sunflower seed, and weed seed saved from the corn shredder. Sunflower seed is particularly good bait for cardinals, cracked acorns and nuts for chickadees and woodpeckers.

If dogs or poultry interfere, fence them out by erecting a temporary fence. If starlings or English sparrows get too thick, thin them out.

Nearby evergreens, vines, or thick bushes help to hold the birds. Discarded Christmas trees are good temporary shelter.

Success in attracting winter birds is largely a question of persistence. If the birds can count on finding food, you can count on their coming after it, and each new year adds more birds of more kinds.

Success in attracting ordinary birds soon whets the appetite for extraordinary ones. A mountain ash in the yard may bring cedar waxwings, or even the rare Bohemian waxwing. A box elder tree may bring evening grosbeaks. Faithful feeding of suet may bring the uncommon redbreasted nuthatch. A south-facing hollow snag wired into the top of a dense evergreen may add the screech owl to your list of guests. A good fencerow, judiciously baited with corn, may lead quail or pheasants out of the fields into your door-yard.

A good feeding station is the best of classrooms for learning ornithology, and one of the luxuries forbidden most city folks. You will enjoy it quite as much as the birds.

Woodlot Wildlife

IN EUROPE, foresters for two centuries tried to clean the woods of every dead, hollow, or defective tree. They succeeded so well that woodpeckers, squirrels, owls, titmice, and other hole-nesting birds have become alarmingly scarce. In Germany, I saw dead oaks laboriously being riddled with auger-holes to encourage woodpeckers.

In Wisconsin, we pay a hunting license to restock the state with raccoons, and at the same time we are chopping down the last hollow trees in which a coon can live. We maintain a closed season on ruffed grouse in the southern woodlots, but grouse are rapidly disappearing from them because the down logs needed for drumming and the brush needed as cover are being removed.

This does not make sense. A few hollow trees, especially durable live basswoods or oaks, and a few dead and down logs are essential to a balanced assortment of wildlife on the farm. The fur-bearers and the squirrels, the rodent-eating owls, the insect-eating woodpeckers, chickadees, and bluebirds are all dependent on dead wood for their breeding places, and hence for their existence.

Hollow apple trees in the orchard, while admittedly not good horticulture, are especially attractive to screech owls, crested fly-catchers, and flickers.

Dead willows in the marsh, after being riddled by woodpeckers, are almost sure "bait" for tree swallows. Dead willows in the southern counties, if over water, may also harbor the beautiful prothonotary warbler.

Wide shallow cavities in creekbank trees are the nesting place of the wood duck.

Tall dead snags near lakeshores always have a chance of becoming an eyrie for the bald eagle or the fishhawk.

If you are lucky enough to have otters on your stream, think twice before you cut a hollow tree with the entrance under water. It may be the "holt" where otter pups are born.

Any hollow tree may become a bee tree. The experience of finding one might help you appreciate that there was a time when no cane or beet sugar was to be had in Wisconsin; one ate honey or maple sugar or went without.

When cutting wood this winter, it is well to remember these manifold uses of defective trees. If a tree bearing a valuable den must be cut, consider saving the valuable part and affixing it to a sound tree. Such artificial dens should face south, and be set so as to shed rain.

On the upper slope of many Wisconsin ravines there are limby crooked bur oaks and white oaks, often with hollow limbs. These veterans grew in the open. They mark the edge of the former prairie. Scars of old prairie fires are imbedded in their stumps. They have escaped cutting because they are crooked and short-boled. Quite aside from their value as den trees, these veterans should be preserved as historical monuments. If your boy can learn to read their history, he will understand better the meaning of his home state and his home farm.

Stories in the Snow

Some Sunday in January when the tracking is good, I like to stroll over my acres and make mental note of the birds and mammals whose sign ought to be there, but isn't. One appreciates what is left only after realizing how much has already disappeared.

Every large Wisconsin woodlot, for example, ought to show the mincing lady-like tracks of ruffed grouse, but few do. There are a dozen counties now grouseless. Why? Because we failed to reserve part of the woods from grazing.

Every woodlot, during winter thaws, ought to show the hurried wanderings of coons emerging hungry from their den trees. Few do, because few woods have any den trees left. The hollow basswoods and white oaks which formerly harbored coons have been chopped out, often by improvident coon-hunters. The same lack of hollow trees has eliminated the flying squirrel, the screech owl, and the barred owl from many a woods.

Out in the corn stubble by the marsh we should find the peculiar tap-dancing tracks of the prairie chicken; instead we find only the trotting-horse stride of a racing pheasant. Why no chickens? Because years ago we plowed up their booming grounds, mowed, burned, or pastured their nesting cover, and then overshot the young in fall. Today we have a dozen chickenless counties, and if fires are not checked in the peat lands, we shall end with a chickenless state.

In the tussock swamp by the tamaracks we can look for a track few people know: the kangaroo-like springs of the jumping mouse. But if the tussocks have been drained, or too hard pastured, the jumping mouse will have disappeared, to be replaced by the prosaic meadow mouse.

In the tamaracks, if you have any, you should find the regurgitated pellets of the long-eared owl. Note well the mouse skulls; three skulls per pellet, one pellet per day, 100 days in the winter, 300 mice per owl per year. Can you afford to let some rabbit-hunter with the trigger-itch shoot him just for fun? Is it worth while to keep a few tamaracks just to have owls around?

By the riverbank, if you have one, there should be, at least at rare intervals, the toboggan-slide of an otter playing in the snow. Most Wisconsin rivers are now otterless; monotonous ribbons of mud and water. A single otter will travel twenty miles of river, and to the mind of the initiated convert that long stretch of mud and water into a personality. In England, otters are common, even in densely settled districts. Why not in Wisconsin?

Winter Care of Plantings

Trees, shrubs, and vines planted for wildlife cover require some forethought and care. Not as much as crops and livestock, but some nevertheless.

Your main winter risk is rabbits. In some years mice may girdle trees; in open winters wandering stock may browse them; in dry, snowless winters, frost may injure them.

Rabbit trouble begins when the young trees have emerged above snow level. It is much worse in cover than in the open. Any small evergreen planted in brushy or weedy ground is liable to be demolished by rabbits. On brushy ground the only defense against rabbits is to surround pine and spruce with cylinders of woven wire, or else to plant such large stock that the rabbits can't reach the terminal shoot. If wire is lacking, loose bundling with straw, weed stalks, fine brush, or burlap answers the same purpose.

I have found that in underplanting woodlots it reduces rabbit damage to place trees away from trails. Rabbits seem to be conventional creatures; they hate to step off the "sidewalk." Screening trees with loose brush makes them less conspicuous and helps cut down damage.

The rabbit appetite for hardwoods likely to be planted as cover varies greatly as between species. The following seem immune to rabbits: grape, Virginia creeper, dogwood, hazelnut, spirea, poplars, honeysuckles. The most eagerly sought hardwood is one almost exterminated from our woodlots and well worth reintroduction: the burning bush or wahoo.

Mice make trouble in plantations only during years of excessive abundance. The years 1935 and 1939 were high in mice. "Highs" occur at four-year intervals, so 1943 is the next in prospect.

Unlike rabbit damage, which occurs above the snow, mouse damage to plantations occurs only below the snow. It is further restricted to ground covered with heavy grass or (less often) weeds. A heavily grassed plantation of small pines may be wiped out by girdling before the owner is aware of the damage. Mouse work can be distinguished from rabbit work by the fact that it is less patchy, and more likely to encircle the stem.

The best defense against mice is clean cultivation. This, of course, implies planting the trees in rows. Where you are threatened with mice but it is impossible or too late to cultivate, you may be able to burn off the grass, first thoroughly wetting down each tree with a sprinkling can. A few singed limbs are preferable to girdled trunks.

All eleventh hour "cures" for rodent pests are of course inferior to natural preventatives, chief among which is a generous stocking of hawks and owls. (Every owl eats from two to five mice per night, and is well worth preserving as a mouser.) In the case of rabbits, the free use of dog and shotgun may also qualify as a natural cure, at least in the eyes of youth.

Do We Want a Woodsless Countryside?

TRAVELERS crossing southern Wisconsin a century ago tell of unbroken prairies where a few decades later nearly every farm had a woodlot. The reason was the invasion of prairie by oaks, particularly red oak, as soon as enough ground was broken to check the free sweep of prairie fires. This invasion was so spectacular, and occurred over such wide areas, that the Wisconsin Agricultural Society debated the puzzle at length, and wondered where the seed came from, and how it got planted.

Today southern Wisconsin is on its way back to "prairie." The reason is clear: drouth and axe take off the old trees; the dairy cow takes off the young ones. The dairy cow eating the tender sprouts off an oak stump or seedling is a more potent deforester than any lumber baron, for the lumberman moves on to pastures new, but the dairy cow doesn't.

To live by dairying, we of course had to make pasture, and the woods was one place to make it. The only question is: *how much* of the woods shall be grazed? The average farm grazes *all.* By grazing all the woods we eventually exterminate woods.

The drouths have greatly speeded the shrinkage of woods, but it takes no forester to see that drouth hits hardest in pastured woods, especially on west edges and south slopes. Unpastured woods seldom show drouth-killed trees.

A woodsless countryside has many disadvantages over and above the obvious one of being hard to look at, and the practical one of forcing the farmer to pay cash for fuel and posts. It is short on wildlife and long on wash and wind. Wash, wind, and wildlife are really community interests. If everybody does his part to get rid of wind and wash and keep wildlife, the community is a better place to live. A thriving woodlot, full of birds, is thus a contribution to the community and a badge of social conduct. Both are rare in this "gimme" world.

Ordinarily any part of the woodlot withdrawn from grazing will regain its thrift and produce young trees to replace the old ones. In some woods, however, the soil has become trampled so hard that young trees can't get started, even when cows are fenced out. In such cases, artificial replanting is the only way out.

I know a farmer who is sawing off limbs in an effort to feed his stove without exterminating his remnant of grazed-out woods. It would have been easier for him to keep the cows out of a small woodlot from the outset. In farming, as in war, it is often hard to retrieve mistakes.

The Farmer and the Rabbit

To the orchardist, truck farmer, or forester a few rabbits are too many.

To the hunter, especially the boy just learning to shoot, lots of rabbits are too few.

Fortunately, the rabbit supply is rather easily manipulated to fit the landowner's wishes.

If you want more rabbits, leave brush, brushpiles, and weed patches. Leave standing corn, or corn in shocks, for winter food. Dense cover and corn will invariably draw cottontails. The corn is even more effective if supplemented by small stacks of alfalfa or clover hay, laid on top of a layer of poles so spaced that rabbits can feed from below.

Another drawing card for rabbits, seldom appreciated by the casual observer, is the woodchuck. The two are allies, not by virtue of treaties, mutual trade, or racial friendship, but by virtue of two common enemies: weather and dogs. The chuck digs holes which the rabbit uses in winter while the chuck sleeps. The rabbit keeps dogs busy in summer which might otherwise dig out the chuck.

This defensive alliance saves many female rabbits in winter, for the females are more inclined to den up underground than the males. If you want to test the truth of this assertion, examine the sex of rabbits in your bag on a zero day. They will be all males; for only males bed out in zero weather.

If you want fewer rabbits, use the gun and beagle, or build some treadle traps. These are far more effective than the old-style box-trap. A working drawing appears in the bulletin "Cottontails in Michigan" published by the Michigan Conservational Department. Traps may be baited with corn ears, old apples, or carrots. Trapping is successful from November to March.

What is "lots of rabbits"? Nobody really knew until scientific census-taking began a few years ago. In one eight-acre marsh in Jefferson County my students trapped and earmarked seventeen rabbits, or two to the acre. In southern Michigan the densest spots showed four an acre. I doubt if the average quarter-section in rabbit country carries more than forty rabbits, and most farms doubtless carry less.

We who have had pines nipped or apple trees girdled by rabbits may, in our irritation, wish there were none. But a rabbitless Wisconsin would be a dreary place. For one thing, it would be devoid of foxes and of the larger hawks and owls, for to them the rabbit is the staff of life. In summer they can turn to mice, but in winter snows there are no mice; they must catch rabbits or starve. The rabbit, then, furnishes the motive power for the horned owl's hooting and for the barred owl's ironic laughter.

Again, it seems likely that some plants eaten by rabbits are, in the long run, benefitted rather than injured. Take sumac, which is severely chewed up nearly every winter. Sumac eventually dies off if not forced to resprout; the rabbit does the forcing. He is the living pruning shears of the wild garden. He snips at random, does lots of good and lots of harm, and is equally unaware of both.

It is a pretty good garden at that.

The Farmer and the Fox

Foxes, in most farming counties, are the sole survivors of Wisconsin's larger wild carnivores. Cougar and lynx are gone, wolf and bobcat are going, but red and gray foxes are still with us, as wild, or wilder, and in some counties just as plentiful, as they were when Marquette's canoes glided down the Wisconsin. The fox, and the old bur oak on the knoll, saw the first wagon-track on the military road; they saw the last fire sweep eastward over the prairies; they saw the first black furrow roll over and point its grass-roots at the sky. The rest of us are, historically speaking, newcomers and greenhorns.

It is important not to cut off these last links with the old Wisconsin. They help remind us whence we came, which sometimes gives hints on whither we should go.

The red fox is tough and needs no coddling; nothing can run him out of his native haunts save den-digging and poisoning. The gray fox, but not the red, tends to disappear where woodlots and

thickets are too heavily pastured. Both species persist longest in those localities offering rock outcrops for ultimate sanctuary. Hunter and hound do not often depopulate a fox range, but foxes may deflate the pride of many a hound, and of not a few hunters.

Most sports have lost value by becoming easy; they are a test of machines rather than of wits. Not so still-hunting for red foxes, Wisconsin style. He who picks up a fox-track on a winter morning, follows it through the night's prowlings, guesses when and where the fox has bedded, stalks the bed, and delivers his shot nose-in-tail, that hunter has performed in a feat worth bragging about. Few ever perform it, and as hunters grow thicker, softer, noisier, and dumber, fewer may even try.

The most fervent friend of the fox will not deny that he takes poultry, nor that when he does, his hide should hang on the barn if the aggrieved poultryman can contrive to hang it there. Whether the mice the fox eats do more damage than the fox does is a moot question; maybe they do; maybe they don't; little is known about what really controls mice.

The ultimate question is whether a foxless, hawkless, owlless countryside is a good countryside to live in. Some of us, for reasons that wholly defy explanation, are sure it is not. Some of us have seen countrysides from which all these really wild things are gone, and in which land-minded people regret their going, and make pathetic efforts to restore them.

The fox is an enthusiastic mouser, especially when mice are plentiful. The red fox, and probably also the gray, kills far more than he can eat, and has lots of fun doing it. The excess mice are "cached," or merely tossed aside. A friend of mine in Michigan followed many red fox trails in the snow and tallied their kills. They got far more meadow mice than deermice; they ate only half of their kill. Shrews and moles, while often caught, were practically never eaten. I have found double-handfuls of partly digested mice thrown up by some fox who had evidently overestimated the capacity of his stomach.

As a rabbit-hunter, the red fox far excels the most active dog of equal size. My Michigan friend found six rabbit kills made by red foxes in a single woods in a single night. Excess prey remains are usually "cached," to be eaten at times when hunting may be poor, or later exhumed by hungry skunks, and finished off by crows, hawks, and owls. Just as Buffalo Bill kept the railroad crews in meat, so the red fox seems to feed the whole hungry tribe of winter flesh-eaters. The difference is that Buffalo Bill (and his like) exterminated their meat supply; the fox does not exterminate anything.

It has long been suspected that foxes hold down wild housecats, possibly by killing kittens. Cats' claws have been found in red fox droppings, but the fox may have picked these up as carrion. Many oddments are occasionally found in fox-droppings: corn, turtle, snake, grass, weasel, bittern, insect, hickory nut, and crow are samples. Does this catholicity reflect appetite, or the national thirst for vitamins?

Foxes, like the wind, are community property; they stay on no one farm. The red fox ranges over a far wider area than the gray. Red fox pups, ear-tagged and released in Iowa, were recaptured during the following winter at points as far as 100 miles away. These pups, however, had been moved from their home range, and like transplanted pheasants, probably wandered abnormally. Tagged pups not moved, i.e., released on the home range of their parents, were recaptured at points less than 20 miles distant.

Fox populations do not increase indefinitely; dispersion, disease, or some other misfortune always trims off the excess. Thus in 1935 grays were overabundant in Wisconsin, but by 1939 their levels had receded to normal. Reds and grays often alternate in dominating a single locality; both species are never simultaneously abundant.

What is meant by abundant? On the Prairie du Sac study area, a high gray fox population in 1935 consisted of 30 on 3,200 acres, or 5 grays per square mile. In 1938 this area had as high as 8 reds on 3,200 acres, but they ranged outside, and probably did not average over 1 per square mile. In Iowa, a census showed 1 red per square mile in one county, and 1 per 5 square miles in another. Grays, in short, may become more abundant than reds before natural processes trim off the excess.

Many of us live out our lives in fox country without ever seeing a fox, or learning anything about his capricious ways. His comings and goings, troubles and pleasures, successes and defeats are veiled in black midnight, and become visible to us only as tracks in the snow. Even the fox-lover, after a life-time of study, is often baffled by his doings. Perhaps his main utility is to knock down our ears, to teach us that there are still some things we know nothing about.

Pines above the Snow

WHEN WINTER has locked up the soil and dispersed the birds and denuded the hardwoods of their leaves, one needs to be assured that things will one day grow again. At such a time a young pine, green above the snow, talks louder than many voices.

The rain falls on the just and the unjust, but pines do not grow wherever rain falls. Only farmers with acid, and preferably sandy, soils are privileged to have real pines. By real I mean pines with that inner bloom which bespeaks abounding health—pines with blue needles, long clean inter nodes, and rifle-barrel leaders aimed straight at the zenith and intending to hit their mark.

I mean, too, *native* pines, white or Norway, sure of their root-hold in native earth—not Scotch, nor Austrian, nor any other stowaway from foreign parts. Nurseries have peddled these foreigners because they are easy to raise. They make a big show for a few years, but they do not stand the test of time.

"Norway" sounds like a foreign pine, but the tree is as native as the ruffed grouse who treads the brown carpet under his branches. He is called Norway because first logged at Norway, Maine.

Pines, like other blessings, come to him who waits. When the labor of planting is done, you wait for a rain. When the plantation is safely rooted, you wait three years for real growth to begin. Then, for a decade or two, you wait all year for May to come, for buds to burst, for waxy "candles" to reach skyward, each year a little farther: first a foot a year, then two feet a year, finally sometimes three feet a year. If, during the pyramiding period, your own clock shows signs of running down, you may gain from your trees a curious transfusion of courage. Pitch, like blood, is thicker than water.

It is a good thing to have more land for pines than time to plant, for thus you plant a part each year, and each year have new trees making their first thrust at infinity.

Your pines, like your children, are interesting to the extent that they are studied minutely. For example: Why did the annual gain in height for 1941 exceed that for 1942? This holds good for almost all plantations, and for wild seedlings as well, throughout the central region.

I don't know the answer. I do know that the drouth year, 1936, registered itself in short growth for 1937. Despite the drouth, the 1936 growth was excellent, reflecting the abundant rains of 1935. Pines, in short, deposit their current paycheck from sun and rain, and pay their bills with savings of the year before. Most other trees live from hand to mouth; they are content to let the Lord provide.

Have you ever noticed that the vigor of a young pine registers in the number of buds at the tip of his leader? Show me a leader with eight buds and I will wager on a long climb skyward when they burst next May. Show me three buds, and I see a weak and ailing tree in need of a pediatrician.

Health also registers in needle-length. Your newly planted pine may have grown three-inch needles during his last year in the nursery, but after the shock of transplantation he may do well to grow needles half as long. At this stage his leader looks like the singed tail of a fox; a year later the new growth will show longer needles, deeper color, and more buds.

Why is a pine called evergreen? Does he keep his needles forever? By no means. A white pine sheds his needles when they are two years old; a Norway pine when they are three years old. Look at your trees to verify this. If you have none to look at, I hope you have something else as good.

The jack pine is a precocious cousin of the real pines. He starts out in an awful hurry, and then loses his wind. He is useful as a pacemaker in a young plantation. Sprinkle a few jacks among your whites or Norways, and they will furnish wind-protection by the time the better pines at last decide that they want to grow up. Wind-protection is an asset to any tree. White pines, unless planted on low moist soil, may fail to perform well without it.

The jack pine, like his human counterpart, is not steadfast. He wants to do something quickly; he doesn't care when or how well. He often puts on two or three separate spurts of growth during a single summer, hence the whorls of his branches do not register his age, as they do in white or Norway. He lays down sappy wood that rots quickly. Altogether he is an amiable four-flusher, but it takes all kinds of trees to make a forest. His usefulness lies in his low standard of living; he plants the green banner on sands too poor to support better trees. After a generation of jacks, such soils may accumulate enough fertility to support Norways, and perhaps, eventually, whites.

Just so did the swaggering coureur de bois prepare the way for the settler, and the settler prepare the way for us.

SPRING

The Farm Pond

I N T H E dust bowl thousands of artificial ponds are being built, with governmental help, by farmers who learned during the drouth to appreciate water. Some day Dakota may build as many ponds as Wisconsin has drained.

The farm pond has many uses: stock water, muskrats, pan fish, water lilies, and last but not least, waterfowl.

Jack Miner, an Ontario farmer, had a small pond behind his barn. Like most ponds, it was duckless, having been "burnt out" by

too much shooting. Jack Miner quit shooting, put out some live decoys and grain. A few wild birds began to drop in. Within a decade ten thousand Canada geese were visiting his refuge each spring and fall. So spectacular a success is, of course, not to be had by all, but any pond, even if temporary, can attract at least a few interesting waterfowl. The time to start is spring.

The first essential is to exclude all shooting. As soon as the ice breaks, put in a few live decoys, preferably of pinioned wild ducks, and some feed, preferably corn. It is illegal to use either live decoys or grain bait on a shooting pond, but on a refuge they are both legal and proper. Avoid too many decoys; they roil the water. If crows and blackbirds are bad, put the feed on a shallow bar under water. If you have no bar, build one by hauling gravel on the ice. If carp prevent placing the feed in the water, feed on an open beach. Diving ducks like bluebill and redhead, however, will not readily find grain unless it is in the water.

Once the ducks start to use your pond, both the numbers and kinds will increase as long as protection is maintained. Some kinds, notably bluewinged teal and mallards, may be induced to stay and nest, especially if a marshy shoreline is fenced against cattle. A protected marshy shore may also attract rails, grebes, coots, gallinules, and bitterns.

One of the most difficult feats is to induce wood ducks to nest in a box. The box is attached to a tree on the wooded banks of a pond or creek. I will supply box designs if you think you have wood ducks and a suitable location for them.

For the greatest variety of bird life, part of the shoreline must be bare of cover. Shorebirds, geese, and most ducks like to loaf where they can see in all directions.

Kingfishers and terns add greatly to the summer bird life of a pond. To attract kingfishers, plant a few dead snags with limbs overhanging the water, and leave a steep bank in the nearest gravel-pit for the birds to excavate their nests. Such a bank will give you hundreds of nesting swallows as a "dividend." To attract terns, build a raft and anchor it in the middle of the pond. Such a safe loafing place is infallible bait for terns and perhaps gulls.

To merely attract birds is only half the game. The next thing is to distinguish the various kinds, to learn the habits, calls, and plumages of each, where they come from and where they go. Some of your callers will be on their way from Carolina to Saskatchewan; others from Patagonia to the Arctic seas. All they ask of you is something to eat and a safe place to wet their feet.

The Marsh

"Shall we fire the marsh?" This is a question which faces the owner of marshlands, especially in the spring following a wet year like 1938. In wet years, marsh hay is likely to be left uncut, and hence must be burned off if the mower is to be used the following summer.

Burning does not hurt a marsh if the soil is wet, and if infrequent enough so that the bushes resprout. But burning which consumes the peat soil is ruinous to all living things. Even light burning, if repeated annually, gradually kills all bush growths, and when the bushes are gone the marsh loses its cover value for birds during deep snows. In hard winters like 1935–1936, quail and pheasants survived only in bushy marshes with feed nearby. All other growths were squashed and buried by snow.

The farmer who prizes his game birds should therefore be careful to keep part of his marsh in bushy growths of dogwood, willow, or elder. If he must burn, he should confine the fire to the area to be cut for hay, keeping it out of the area to be left as bush cover. The burning and grazing of marsh cover is subtracting from Wisconsin's game bird supply faster than the State Game Farm is adding to it.

Here is an actual case—in 1933, I met a farmer burning a peat pothole "to get rid of the ragweeds." I asked if he had burned before. Yes, he burned every year. I persuaded him that perhaps fire was the cause of the ragweeds. He agreed to try the idea, and has not burned since. By 1937, the ragweeds had disappeared, having been replaced by aster, goldenrod, dogwood, and elder. A feeding station was maintained in this pothole every winter. While it was still in ragweed (the giant variety), not a bird could be persuaded to use the station. As the more varied vegetation gained a foothold, the game population began to build up. Last winter this eight-acre pothole carried 28 pheasants, 20 Hungarian partridges, and 21 cottontails.

Marsh wildflowers as well as marsh game birds suffer from too much fire. Most of the ladyslippers, pitcher plants, and other bog flowers thrive only under the shade of tamaracks. Repeated marsh fires push back the tamaracks until they disappear. Some ladyslippers require live sphagnum moss to grow in, and hence are destroyed by fire. Most Wisconsin marshes have already lost all their tamaracks, and with them most of their bog flowers, through the combined action of fire and grazing.

Curiously enough, some birds require freshly burned marsh for nesting and feeding. The Brewer's blackbird, for example, nests in Wisconsin only in freshly burned cattails. Jacksnipe and geese like a burned marsh for feeding during the spring migration. A few valuable bird foods, especially the false climbing buckwheat, are greatly increased after light-burning a peat marsh. Needless to say, if fire is to be used at all it should be used early, before nests are built, and in no case should a marsh be burned deeply or without being sure of where and how fire will help. There is truth in the old saying: "Fire is a good servant, but a bad master."

Evergreens for Cover

YOUNG EVERGREENS furnish more shelter per square rod, for a greater variety of wildlife, than any other vegetation. The farmer who wants to hold his birds in winter, but has only a few odd corners to devote to cover, will do well to plant evergreens on them.

It takes five years, however, for small evergreens to reach a size useful as wildlife cover. Failure is probable if maintenance is neglected, or if the kind of evergreen does not fit the kind of soil.

On any limey soil, any soil which will grow alfalfa or clover without liming, I recommend red cedar and creeping juniper for dry locations, white cedar for moist locations. On acid soil use white pine, white spruce, or Norway spruce for moist situations; red (Norway) pine, jack pine, or Scotch pine for dry situations.

White pine, spruces, and white cedar do best in partial shade. Red pine and jack pine tolerate little or no shade. If in doubt about what kind of evergreen to plant, select the kind native to the locality.

Use nothing smaller than 2-2 stock, i.e., four-year-old trees which have grown two years in the seedbed and two years in the transplant bed. In heavy grass or where rabbits are thick, even larger stock is desirable. White pines and spruces may be wiped out by rabbits unless the stock is large enough for the tip to be out of reach. Red cedar is rabbit-proof; jack pine is proof against cottontails, but not snowshoes.

If you must use stock smaller than 2-2, put it in the garden and let it grow until at least a foot high.

If you plant white pine, make sure your farm is free of gooseberries and currants which carry the white pine blister disease. Do not plant red cedar near apples. These two species share a rust disease which may destroy both.

The worst enemy of evergreen plantings is grass. If the planting site is sodded, I advise clean cultivation for a space of three feet from each tree before planting, and for two years afterward. Herbs or weeds, if not too rank, do no harm, and in the case of the shade-tolerant evergreens, may be beneficial.

On ground which contains vestiges of grass, especially quack, do not mulch or else mulch heavily. A light mulch merely protects the grass while it forms a sod.

Land frequently flooded by overflow is not suitable for planting evergreens. White pine, spruces, and white cedar tolerate short periods of flooding, but the other pines do not, especially if flooded in summer.

All plantings in pastures must be fenced, with the possible exception of red cedar. Red cedar resists grazing, but may be destroyed by running.

To make evergreens valuable to wildlife, the design and location of the plantings is important. Cold windy locations are not worth planting. If you have a warm south-facing bank, use evergreens as a windbreak to make it warmer. On such banks the drouth resistant species like red cedar, creeping juniper, red pine, or jack pine are best. Clumps or lines of evergreens connecting marsh or wood-lot cover with feeding grounds are desirable. Clumps or lines of evergreens are often a successful means of leading wildlife to a feeding station in the farmyard.

Evergreens planted for wildlife should not be pruned. Their value as winter cover lies in the low-hanging branches which sweep the ground. As the trees grow older these low branches die, and should be replaced by a new planting of young trees.

Nearly all wildlife species which winter in Wisconsin make use of evergreens. Pheasants, quail, and Hungarian partridges resort to them during blizzards. Cardinals, chickadees, juncos, tree sparrows, bluejays, cedar waxwings, and redpolls can be held without them, but they winter in greater numbers where evergreens occur. Long-eared owls, evening grosbeaks, crossbills, and pine siskins seldom winter where evergreens are absent. Wintering bluebirds and robins are especially attracted by red cedar berries; crossbills by pine or spruce cones.

There are localities in Wisconsin where no evergreen will fit. Thus there are soils too limey for the acid-loving species, too near apples to risk red cedar, and too dry for white cedar. In such localities it is better to use grape tangles for wildlife cover.

When the Geese Return

ONE SWALLOW does not make a summer, but one flock of honkers, winging northward through a murky March thaw, make a spring, come what later blizzards to the contrary notwithstanding.

What chance has a farmer to induce the migrant flocks to settle down and stay awhile? This is a practical question in wildlife conservation. The future of geese is largely a question of hospitable farmers.

If you happen to live on one of the historic "goose prairies," your chances are very good. Geese from time immemorial have watered at certain Wisconsin lakes and fed on the nearest large prairie. What they ate in the days before corn came is a puzzle, because we know so little about what plants covered the original prairie. Probably they ate the seeds of wild legumes and the bulbs of nutgrass. Today they eat corn and the leaves of winter wheat or rye.

Granted you are in or near a goose prairie, what are the requirements for attracting geese? Mainly a large bare expanse of stubble offering corn and winter grain, absolute protection, and if possible, live decoys. Live decoys for shooting are now illegal, but for "baiting" a refuge decoys are both legal and effective.

If you can muster this combination, and have patience, you will eventually attract geese.

By protection I mean complete freedom from shooting over a period of years. Geese have a long memory. Several neighbors who pool their efforts have a better chance to ban the hunter than one farmer acting alone.

A farm pond with bare shores is an additional inducement, for the geese can then dispense with their daily trip to water at a lake. The pond should contain gravel, but this occurs naturally in most Wisconsin ponds. If it offers a gravelly island, barely awash, it is ideal. Deep ponds with wooded shores and no islands or bars are unfavorable for geese.

It is astonishing that more Wisconsin farmers have not built themselves a goose-show. Once your reputation is established, the geese will pile in, spring and fall. One Canadian farmer (Jack Miner) got so many geese that the government had to chip in to help with the feed bills. If you want geese, now is the time to advertise the advantages of your farm.

If we plan to increase numbers and to attract greater numbers of these noble birds here, we shall have to do some planning and a little work. They remember feeding grounds that furnish what they can eat. They also remember the place where they have had a scare.

Bur Oak: Badge of Wisconsin

WHEN SCHOOL children vote on a state bird, flower, or tree, they are not making a decision; they are merely ratifying history.

When the prairie grasses first gained possession of our southern counties, they thereby decided that the characteristic tree of this region would be the bur oak, for the bur oak is the only tree that can stand up to a prairie fire and live.

Have you ever wondered why that thick crust of corky bark covers the whole tree, even to the smallest twigs? This cork is armor. Bur oaks were the shock troops sent by the invading forest to storm the prairie; fire is what they had to fight. Engineers didn't discover insulation; they copied it from these old soldiers of the prairie war.

Botanists can read the story of that war for twenty thousand years. The record consists partly of pollen grains embedded in peats, partly of relic plants "interned" in the rear of the battle, and there forgotten. The record shows that the forest front at times retreated almost to Lake Superior; at times it advanced far to the south, for at times spruce and other "rear guard" species grew beyond our southern borders. But the average advancement of the forest was about what it is now; and the outcome of the battle was a draw.

One reason for this was that there were allies which threw their support first to one side, then to the other. Thus rabbits and mice ate the prairie herbs in summer and girdled the oak seedlings in winter. Squirrels planted acorns in fall and ate them all the year. June beetles undermined the sod in their grub stage, but defoliated the oaks in their adult stage. But for this geeing and hawing of allies and hence of the victory, we should not have today that rich mosaic of prairie and forest soils which looks so pretty on a map.

In the 1840's a new animal intervened: the settler. He didn't mean to, he just plowed enough fields to deprive the prairie of its immemorial weapon, fire. A rout followed. The oaks romped over the prairie in legions, and "overnight" what had been the prairie region became a region of woodlot farms. If you doubt this story, go count the rings on any set of stumps on any "ridge" woodlot in southwestern Wisconsin. All the trees except the oldest veterans date back to the 1850's and the 1860's, and this was when fires ceased on the prairie.

Thus, he who owns a veteran bur oak owns more than a tree. He owns an historical library, and a reserved seat in the theatre of evolution. To the discerning eye, his farm is labeled with the badge and symbol of Wisconsin.

Bluebirds Welcome

To NOTE the arrival of the first bluebird, like tapping the sugar bush, is an authentic ritual of spring. Most farms, however, are content to let the bluebirds arrive, and depart, without offering them a place to stay.

In the old days when every farm had hollow apple trees and wooden posts full of woodpecker holes, there was no need to provide housing for bluebirds. But today the hollow apples are gone, and the wooden posts are going. The more "modern" the farm, the greater the need for bluebird houses. I once tallied 100 farms and found that only 12 had bird houses of any kind.

Bluebirds once nested in towns and villages, as well as in open country. English sparrows and starlings have completely routed them from urban habitats, and are now by way of routing them from farmyards as well. Hence an accurate .22 rifle is a good tool for rebuilding bluebird prosperity. The trouble with the rifle is that it may be turned against hawks and owls, or other birds just as desirable as bluebirds.

The rifle is not the only way to foil starlings. One very simple way is to erect bluebird houses not over eight feet above the ground. Starlings will not nest at such low levels, while bluebirds prefer to.

Bluebird houses may be built of old lumber, but a better-looking house may be made by ransacking the woodpile for hollow sections. Most woodpiles contain hollow cylinders of convenient length (6 to 12 inches), with hollows 4 to 8 inches in diameter. Bore, chop, or saw an entrance hole in such a cylinder, tack on a top and bottom board, and your bluebird house is complete. Set it on a high fence post, or on the top of a short pole set in the ground. Place the house in a fencerow or open pasture, never in dense woods. Have it ready by April 1, because bluebirds like to spend several weeks at house-hunting, and they lay eggs by May.

All bird houses should be built so they can be opened and cleaned yearly. A convenient way is to use short nails for the roof-board, so it can be knocked off easily and put in place.

A still more natural way to accommodate bluebirds is to leave rotten stubs or limbs for the woodpeckers to riddle with holes. Bluebirds will use the old holes of the larger woodpeckers. A dead willow, aspen, soft maple, or elm becomes workable sooner than the more durable oaks, but by the same token it lasts fewer years before toppling down. A really conservation-minded farmer never cleans up all his dead trees, for by doing so he evicts his bluebirds, woodpeckers, and flying squirrels.

The spring of 1941 is particularly suitable for starting bluebird houses, for the reason that bluebirds "took an awful beating" in the extraordinary storms which swept over the gulf states during the winter of 1939–1940. Bluebirds were very scarce last summer, and need all the help we can give them to recover their normal abundance. This wholesale killing of bluebirds by winter or spring storms is repeated at intervals; the worst killing came in 1894, and is known historically among old-timers as "the bluebird storm."

Back from the Argentine

WHEN DANDELIONS have set the mark of May on Wisconsin pastures, it is time to listen for the final proof of spring. Sit down on a tussock, cock your ears at the sky, dial out the bedlam of meadowlarks and redwings, and soon you may hear it: the flight-song of the upland plover, just now back from the Argentine.

If your eyes are strong, you may search the sky and see him, wings aquiver, circling among the woolly clouds. If your eyes are weak, don't try it; just watch the fence posts. Soon a flash of silver will tell you on which post the plover has alighted and folded his long wings. Whoever invented the word "grace" must have just seen the wing-folding of the plover.

There he sits; his whole being says it's your next move to absent yourself from his domain. The county records may allege that you own this pasture, but the plover airily rules out such trivial legalities. He has just flown 4,000 miles to reassert the title he got from the Indians, and until the young plovers are a-wing, this pasture is his, and none may trespass without his protest.

Somewhere nearby the hen plover is brooding the four large pointed eggs, which will shortly hatch four precocious chicks. From the moment their down is dry, they scamper through the grass like mice on stilts, quite able to elude your clumsy efforts to catch one. At thirty days the chicks are full-grown; no other fowl develops with equal speed. By August they have graduated from flying school, and on cool August nights you can hear their whistled signals as they set wing for the pampas, to prove again the age-old unity of the Americas. Hemisphere solidarity is new among statesmen, but not among the feathered navies of the sky.

The upland plover fits easily into the agricultural countryside. He follows the black-and-white buffalo which now pasture his prairies, and finds them an acceptable substitute for brown ones. He nests in hayfields as well as pastures, but unlike the clumsy pheasant, does not get caught in hay mowers; well before the hay is ready to cut, the young plovers are a-wing and away. In farm country, the plover has only two real enemies: the gully and the drainage ditch. Perhaps we shall one day find these are our enemies, too.

There was a time, in the early 1900's, when the Wisconsin prairies nearly lost their immemorial timepiece; when May pastures greened in silence, and August nights brought no whistled reminder of impending fall. Universal gunpowder, plus the lure of plover-on-toast for post-Victorian banquets, had taken too great a toll. The belated protection of the federal migratory bird laws came just in time. Today the eight pairs of plovers which in 1935 nested on the Faville Grove area, in Jefferson County, have increased to twenty-two pairs. Plovers thrive on many of the cow-built prairies of the dairy belt. Plovers have invaded the fire-built prairies of the pine belt. Progress has granted a stay of execution. If Progress can now grant sharper ears for things American to young America, the race of plovers may yet survive.

Sky Dance of Spring

I owned a farm for two years before becoming aware that the sky dance is to be seen in my woodlot, each evening in April and May. Once discovered, my family and I are reluctant to miss even a single performance.

The show begins on the first warm evening in late March, at exactly 6:45 P.M. Station yourself near an opening in woods or brush bordering a marsh, and listen. Soon you will hear the overture, a queer note sounding a little like a hoarse frog and much like the summer call of the night hawk: peent-peent-peent, monotonously repeated. Once you hear the peent, move up cautiously to get the performer between you and the western sky.

Suddenly the peenting ceases and a bird climbs skyward in a series of wide spirals, emitting a musical twitter. It is the male woodcock, in mating display. Up and up he goes, the spirals steeper and smaller, the twittering louder and louder, until the performer is only a speck in the sky. Then, without warning, he tumbles like a

crippled plane, giving voice in a soft liquid warble which even a March bluebird might envy. At a few feet from the ground he levels off and returns to his peenting ground, usually to the exact spot where the performance began. The spot is always in an opening; a pasture, a haymeadow, or a bare rock, and always in or near woods or brush. The sky dance is repeated dozens of times each evening, and by moonlight it lasts the night.

This little drama is enacted nightly by at least thirty pairs of woodcocks within a mile of the outskirts of Madison, i.e., within a mile of thousands of people who sigh for dramatic entertainment. It is enacted nightly in the woodlots of thousands of farmers who seek the better life, but who harbor the illusion that it grows in department stores and theatres. They are unaware that part of it grows on the back forty. There are less than a hundred people in Wisconsin who know, and annually enjoy, the sky dance.

The woodcock is living refutation of the theory that the main utility of game birds is to serve as a target, or to pose gracefully on a slice of toast. No one would rather hunt woodcock in October than I, but since learning the sky dance I find myself calling one or two birds enough.

The nearly universal grazing of woodlots and drainage of marshes is fast evicting the woodcock from southern Wisconsin. Light grazing improves woodcock cover by providing openings for the sky dance, but the kind of grazing which removes all the brush and young timber is ruinous. An ideal woodcock range consists of a springy alder or dogwood swamp adjoining spotty thickets of hazelbrush, blackberry, young popples, and young oaks. On such a range I have counted as many as twenty sky-dancing pairs per square mile, but this is exceptional. Overgrazing, overcutting, and drainage have already cleaned many a township of its last woodcock, without a single person being aware of the loss.

Woodcocks are just now in special need of help, for the spring blizzards of 1940 caught them too far north; thousands froze and starved. Such natural losses are unavoidable, and would do only temporary damage if the breeding ranges were in good shape. But in the dairy belt the breeding ranges are being improved to death. More people should learn the sky dance; we cannot conserve what we do not know exists.

Farmers and Ducks

A FARMER WHO happened to adjoin a golf links turned an honest penny by selling muck for dressing greens. When the borrow-pit he had dug filled with water next spring, he was surprised to find it occupied by a pair of bluewinged teals. The water dried up, but not before the teals had brought off a brood.

Banding returns show that when ducks once nest in a given spot, they tend to return to it year after year if the water is still there. It is the hen who returns, not the drake or the brood. If, then, a duck-minded farmer can once induce ducks to nest, he is set for future years, provided his hens can survive the gauntlet of the guns.

Wisconsin now breeds fewer ducks than formerly because fewer Wisconsin hens run the gauntlet, especially on opening day when most of the ducks are "locals." When the mob arrives at daylight on October 1, the duck-minded farmer will do well to say "no."

A poet once said, "Each man kills the thing he loves." Many a duck-loving farmer unwittingly injures the wild flights by maintaining scrub mallards on his pond, and allowing them to interbreed with wild mallards. Birds differ in their capacity to resume wild existence after a period of domestication. Mallards and wild turkeys gradually become unfit to survive in the wild; pheasants do not. When barnyard mallards cross with wild ones, they pull the resulting progeny down to lower levels of fitness.

What is fitness? Mainly, no doubt, a certain pattern of instinctive behavior, but there are also physical characters which denote scrub stock, and these characters are easily recognized. Among them are pot-bellied posture, short bill, heavy leg, and a tendency to nest abnormally early. Any kind of color aberration denotes an advanced degree of unfitness. If you can't prevent your scrub flock from mixing with wild mallards, it helps some to cull out the worst aberrations, and let them adorn the platter instead of the pond.

In Canada, in mild falls, a considerable number of mallards fail to leave the breeding marshes; as cold weather comes on they get thin, and then die in the first blizzard. Such non-migrants consist partly of cripples; they probably consist largely of normal ducks whose migratory urge has "gone by"; they may consist in part of scrubs whose migratory instincts were defective to begin with. In any event, it is a doubtful favor to wild duckdom to feed such unfit birds, or to ship them south.

The crying need of duckdom at this time is thousands of farm ponds and marshes where ducks can feed and rest secure from guns. The abuses on shooting grounds, both public and private, are growing far faster than the refuges where no shooting is allowed. Governmental and state refuges cannot possibly spread far enough or fast enough to offset the decline in duck-hunting ethics, or the increase in reckless use of rapid-fire shotguns.

Cliff Swallows to Order

How would you like to have a thousand brilliantly colored cliff swallows keeping house in the eaves of your barn, and gobbling up insects over your farm at the rate of 100,000 per day? There are many Wisconsin farmsteads where such a swallow-show is a distinct possibility.

Cory Bodeman, of Deerfield (Dane County), had 1 pair in 1904; today he has 2,000 pairs. This phenomenal increase is undoubtedly the direct result of his painstaking care for the welfare of his birds. Mr. Bodeman has a recipe for swallow management.

First, you need a rough surface to which the mud nests can be attached. This, on a barn, means unpainted rough lumber. If your barn is already painted, you can tack on a rough board under the eaves, where the rains cannot loosen the nests. On Mr. Bodeman's barn he has added a 1" x 2" horizontal rail, nailed over the vertical battens, to give his nests additional support.

Second, there should be a colony of cliff swallows in the neighborhood, from which a new colony can be derived. Mr. Bodeman's birds are constantly colonizing new barns as far as two miles from his barn.

Third, you must shoot off the English sparrows which build their nests in the old swallow nests before the swallows arrive, and which pester them continuously throughout their nesting season. Mr. Bodeman kills as high as 600 English sparrows each spring, using .22 calibre shot cartridges.

Fourth, you should leave a part of last year's nests in place through April to furnish shelter for the newly arrived swallows in case of a snowstorm or a cold rain. But *you must tear down all old nests by May 10,* and if possible, burn their contents, in order to control the parasites which winter in the old nests, and which are constantly carried about by English sparrows. Parasites will kill the young swallows if they are allowed to increase.

Fifth, you should provide trickles of water during dry springs to provide mud for the nest-building. A leaky watering trough in a barnyard is a good source of mud. The season for nest-building is about May 15.

Cliff swallows usually arrive in late April and begin building nests soon after arrival. The young are being fed during haying season, when the old birds often follow the mower to gather insects for their broods. They have raised their young and dispersed by July 20. They migrate south in September, and spend the winter in Brazil and Argentina.

The species is growing scarce, due no doubt to too many English sparrows and too much paint on barns. In addition, some farmers deliberately destroy swallow nests in the erroneous belief that the swallow parasites spread to people and livestock. Bird parasites are adapted to live only on birds, and cannot spread to livestock or people. This old prejudice is an inheritance from the dark ages, and reflects discredit on the culture and education of those who still give it credence.

The cliff swallow originally nested on cliffs. Its use of barn-eaves as a substitute for cliffs is a habit acquired since the settlement of the country. Paint and English sparrows may eventually make it a rare bird in prairie regions, unless a substantial number of farmers learn to take an interest in its welfare.

SUMMER

Wildflower Corners

WISCONSIN wildflowers are of three groups: the prairie flowers, the woods flowers, and the bog flowers. Each group requires a distinctive habitat, each group responds differently to grazing, mowing, picking, and burning, hence each group has its own distinctive requirements for conservation.

The prairie group, for example, is not injured by fire, provided the fire comes before or after the growing season. Fire, in fact, may be beneficial to prairie flowers in preventing trees and brush from shading them out. Grazing, however, is fatal. We do not know why a cow in a pasture eradicates plants when the free-coming buffalo and elk did not. Perhaps a free-roaming cow not confined by a fence would be less thorough in her picking, and hence less deadly.

Prairie plants can stand mowing if not repeated too often. One of the best ways to preserve prairie flowers in a wild haymeadow is to reserve an unmowed strip each year, rotating the location of the strip. This enables each plant to go to seed occasionally, and incidentally improves cover for prairie chickens and pheasants.

Remnants of prairie vegetation occur on ungrazed roadsides as well as in haymeadows. The best conservation method in such spots is to burn early, mow late, and never graze.

The prairie flowers, in the order of their blooming, include pasque flower, white and small yellow ladyslipper, blazing star, prairie clover, butterfly weed, compass plant, ladies' tresses, and blue aster. Of these, the ladies' tresses and ladyslippers grow only on marshy prairie. The remainder are upland prairie species.

This group, in the order of blooming, includes bloodroot, hepatica, windflower, Dutchman's breeches, jack-in-the-pulpit, ginseng, white, nodding, and red trillium, blue phlox, and large yellow ladyslipper. Many ferns have the same habitat requirements, and may be grouped with the woods flowers.

The woods flowers are disappearing rapidly from grazed woodlots, partly because of direct damage by grazing, but especially because of the combination of grazing, timber cutting, and trampling. These changes admit light and thus admit nettles, burdocks, and finally grass. The woods flowers cannot compete with weeds and grass.

This group includes the pink moccasin flower, pitcher plant, grass pink, rose pogonia, bog laurel, and bog rosemary.

All these bog species require moist acid peat for survival. Drainage and grazing are fatal because they admit peat fires, weeds, and grass.

Conservation of bog flowers is a matter of reserving part of the bog against drainage, grazing, cutting, and burning. The earmark of a healthy bog is green moss. Give the bog flowers green live moss and protection from excessive picking, and they conserve themselves.

Wildflower corners are easy to maintain, but once gone, they are hard to rebuild.

Windbreaks

FARMERS LEARN by experience, but in matters affecting conservation they sometimes learn pretty late.

In the 1920's farmers pulled all the hedges out of the cornbelt.

In the 1930's under the name of "shelterbelt," the government replanted hedges all over the dust bowl. The time may come, given wind and drouth enough, when we shall replant them in the cornbelt. We have already had wind and drouth enough in the sandy counties of Wisconsin; hundreds of miles of pine windbreaks are beginning to line the highways, farm boundaries, and field fences.

Windbreaks are good or bad depending on one's style of thinking. If one thinks as a "lone wolf" they are bad, for they use up good land. Why not let the neighbors stop the wind before it gets to your place?

But if one thinks as a social animal they are good, for with too much wind all the land may become bad, as it did in the dust bowl. Why not help the neighbors stop the winds?

If windbreaks become general, they will have a large effect on wildlife. Scores of species of birds and mammals will have new nesting and wintering cover. Those which do not want or need cover will have plenty of open ground left.

How shall we reconcile this new enthusiasm toward windbreaks with the old hostility toward fencerows? A fencerow is a natural windbreak which springs up without cost. If kept under reasonable control (by cutting at intervals), it is nearly as effective as the artificial windbreaks which eat up cash and labor. Perhaps we are due for a change of attitude toward fencerows. They use up land and sometimes harbor insects, but if we abolish them we lose our birds and increase our wind.

The Farm Arboretum

Most of the original inhabitants of Wisconsin have been run out of the country.

This applies not only to Indians, but also to grasses, flowers, shrubs, and, in part, to trees. Birds and mammals have retained a high proportion of native species. Of the hundred common birds we see on the farm, only three are "foreigners."

The farm landscape, with the exception of the trees in the woodlot, consists largely of foreign plants. All our grains except corn, most of our grasses and hays and weeds, all of our orchards and vineyards, most of our vegetables, and most of our ornamentals, "do not belong."

The city landscape shows the same predominance of exotics. Wisconsin, home of the "sugar bush," has surrounded its capitol with European maples.

Pests and diseases are working year by year further to reduce the native plants in the farm landscape, and to replace them with "weedy" species, often of foreign origin. Thus the white grub is killing our bur oaks and white oaks, the blister rust threatens our white pines, the sawfly our tamaracks, and a new disease is about to overtake our elms.

The term "farmer" means one who determines the plants and animals with which he lives. Farming is the most visible distinction between men and animals. There are said to be other distinctions, but they are not so visible.

The farmer, then, might be presumed to have a historical interest in preserving samples of the plants which originally covered Wisconsin. Preserving native prairie plants, bog plants, and woodlot wildflowers has been discussed in other issues. This is about trees and shrubs, especially the rarer ones not usually encountered either in woodlots or farmyards.

A few farms in the southern counties have native hackberries, sycamores, or Kentucky coffee trees. These are worth preserving as the northernmost outposts of these species. By learning the significance of the lone sycamore on his creek bank, the farm boy is bound to gain a truer picture of the whole plant and animal kingdom, and to fill his mind with those unanswerable questions, the presence of which is the true criterion of education. Why, for example, do most northernmost outposts of trees occur along rivers? How did the seeds get transported upstream? Did some Indian child play with a sycamore button, and then toss it out of a northbound canoe?

Other outposts are quite as interesting. Who owns the southernmost tamarack, white cedar, black spruce, hemlock, or ground yew? Who has the westernmost beech tree? Are there any native redbuds in Wisconsin?

Some of our rare natives are of startling beauty. Have you seen a native wahoo or burning bush when the fruits glow in the October sun? Just why should all the wahoos planted on the university campus be of the Asiatic species? Is education something accomplished only on blackboards?

These are my suggestions for the farm arboretum: Tolerate foreign trees when they behave, but admit only natives to full citizenship. Let the arboretum be not a single segregated spot, but all odd corners where the soil fits the needs of some tree. Attach ten times more value to the tree that came in on his own steam than to the tree you had to plant. Label each notable tree, not with pieces of tin or wood nailed to the bark, but with pieces of thought and understanding nailed to your mind. Make the rounds every month of the year and see to your charges. Trees, like humans, thrive on being looked at.

Pheasant Planting

Planting pheasants, like planting seed, may yield a good crop or none at all, depending on the skill and care used. Much is known about how to raise pheasants; little about how to plant them for high survival.

First of all, the planting stock must be right. Pheasants less than eight weeks old seldom survive, and from then on survival gets better with age, up to full maturity. Underdeveloped pheasants survive poorly at any age. An eight-week-old cock should, if well developed, weigh fourteen ounces. Pheasants which have learned foraging in roomy pens survive better than those from crowded ones.

Second, the method of planting must be right. Pheasants which learn to wander gradually from the pen "go wild" better than those dumped suddenly and violently from pen to covert. Violent releases are known to lose weight, and if this loss is severe the bird may die.

Third, the range must be right if the planting is to "stay put." It is hard to hold pheasants on bare uplands, especially after frost and shooting begins. The better the food and cover and the lower the land, the easier it is to hold birds through the fall and winter. In really severe winters it is impossible to hold pheasants except on bushy marsh, that is, marsh containing spots where willows and dogwoods have gone ungrazed and unburned.

Farms which lose their pheasants in winter because they lack marsh cover may regain them in spring, especially if there are fence-rows, ungrazed woods, and grassy corners suitable for nesting. Such summer ranges must depend on some neighbor within three miles furnishing the winter marsh. Returns from banded pheasants show that the birds travel at least that far to get to good winter range.

Shooting hastens the downhill movement in fall. The farmer may reduce the shooting exodus by shooting only in moderation, by keeping both dogs and guns out of the best spots of cover, and by feeding in those spots. For such "bait" feeding, patches of stand-

ing grain are best. Corn, sweet corn, buckwheat, and sunflowers are good "baits" to hold fall birds.

On range already well stocked with wild pheasants, it is doubtful if much is gained by planting more. Artificially reared birds are at first excluded from wild pheasant "society," as shown by the fact that marked plantings do not at first appear in wild flocks. They are pushed out, and if the wild birds are abundant, the artificial birds may be pushed off the farm.

If in doubt about the survival of your artificially reared pheasants, why not band them and see how many show up in the bag this fall, and where? You may have a surprise coming. By using colored bands, you can follow the dispersal from the planting spot. By weighing the bagged birds, you can also see whether your artificially reared birds have developed as well as the wild ones. There are no universal rules for pheasant management, and each landholder will do well to study his own problems. Banding is one of the surest and easiest ways to find the facts.

Pheasant Damage

MANY A farmer is glad to have a few pheasants around but is worried about what many pheasants might do to his corn.

Corn pulling seems to start with crowding. It began on the university farms when the pheasants increased to the level of one bird per two acres. Once started, we had to cut our pheasants down to a fourth of their former numbers before getting relief. Moral: Don't let pheasants pile up to the danger point.

Most Wisconsin farms are far below the danger point, and could support many more birds without risk.

Pheasants thick enough to damage corn are also thick enough

to crowd out the more valuable native birds, such as quail and prairie chicken, neither of which is known to damage any crop.

Corn pulling occurs in May when all the pheasants are in "harems" and some hens already have nests. The more widely scattered the nesting cover, the less the risk of damage. Fencerows, small patches of weeds or brush, evergreen snowbreaks, ungrazed woodlots, and ungrazed stream banks all help to scatter pheasants at corn-pulling time.

Corn pulling done by gophers and blackbirds is often blamed on pheasants. Even an expert cannot always distinguish the work of one from that of another. The best anti-gopher insurance is a pair of nesting red-tailed hawks in the woodlot. There is no premium coming due on such a policy.

Scattering corn on the surface is often recommended as a remedy for corn pulling. I have tried this, but the pheasants preferred the sprouted kernels in the ground.

In dry summers pheasants often peck tomatoes, melons, and grapes, evidently in search of water. Pans of water are said to give relief, but are laborious to maintain. The best automatic thirst-quencher for birds is a mulberry tree, planted in the open near the orchard or garden. Such a mulberry is a great attraction for songbirds as well as game birds, and is good for poultry as well. Mulberries standing on fertile ground seem to bear the largest crops. Shaded trees or those on heavy sod often fail to bear at all.

Next to mulberry, the best thirst-quencher for birds is the native black cherry, but the season of bearing is shorter.

The damage done by birds is like that done by dogs or children; if you like them well enough there are ways to get along.

Bird Houses

PUTTING UP bird houses is one of the simplest and easiest ways to get more pleasure out of farm life. Why do so few farmers avail themselves of it?

I once counted bird houses visible from the highway and found that only 12 farmsteads out of 100 had bird houses. Did the owners of the other 88 have so much pleasure already that they needed no more? Or were they so preoccupied with big troubles as to forget about small birds? I'll leave you to guess.

The Indians, even before the arrival of the white man, put up gourds for martins. We are told this in a little publication which is scholarly, warm, and human, despite its being a government bulletin: Farmers Bulletin 1456, "Homes for Birds." In it is found all the information one needs about dimensions, materials, and locations of houses for the species of birds which ordinarily accept man-made quarters.

The advanced "bird houser" may wish to try his hand with some of the more difficult species. No one, for instance, has yet induced the horned owl or the barred owl to accept artificial quarters. Natural hollow snags suitable for these large owls are getting scarcer and scarcer in Wisconsin woodlots. Whoever succeeds in getting them to accept an artificial "snag" may plume himself on the achievement.

If you have a stream or pond, you have a chance to induce wood ducks to nest. Specifications for wood duck houses have been greatly improved during the last two years by the pioneer work of the Illinois Natural History Survey. These specifications are available from the Survey (at Urbana, Illinois) in a popular bulletin.

If you have water in your woods and live in the southern counties, you might want to try for that jewel of all the woods birds, the prothonotary warbler. No one has ever "domesticated" him. He nests in old woodpecker holes in decayed willows standing in water. If you have none, it is easy to offer an imitation of this set-up.

It is a short step from the erection of artificial quarters to the conservation or creation of natural ones. When the highway department leaves a vertical bank in a gravel pit or highway cut, the bank swallows and kingfishers promptly take possession. Why this passion for grading all such banks to the "angle of repose"? A graded bank is useless to the birds.

I knew a farmer who took great pride in a colony of cliff swallows which plastered their mud nests on the vertical boards under the eaves of his barn. When it came time to paint he was in a quandary; paint would make it impossible for the mud nests to adhere to the wood. He solved the problem neatly by painting all but the upper foot or two of his walls. Without this ingenious compromise he would have killed more than two birds with one paint brush.

When trees die in the woodlot it is the rule of "good" farmers to cut down the snags. Why not leave a few for woodpeckers, flying squirrels, and hawks? In Europe, where the manicuring of wood-

lots was once carried to absurd extremes, it is now the sign of a "progressive" owner to see a few snags on the skyline of woodlots or fencerow.

If you want bush-nesting catbirds and threshers right at your window, but your shrubbery is not dense enough to meet their specifications for a nest site, try tying bundles of stems together loosely with a cord or wire. The response is often magical. This trick, and many others in the same vein, was first used by Baron von Berlepsch in Germany fifty years ago. Germans once had time to think of such things.

Wildlife and Water

CONTRARY TO popular supposition, it is not necessary for most wildlife species to drink water. We know this because many kinds live and thrive on waterless range, where they get their water from juicy insects, fruits, plants, and dewdrops.

Most kinds of wildlife will, however, drink and bathe if they are offered the chance, and prefer to live on well-watered range. The farm which includes permanent springs, streams, and ponds is therefore better wildlife range than the farm without natural water, especially during dewless drouths when there are few fruits or succulent shoots. The chicks of all upland game birds are known to drink dewdrops, and dewless periods may be very damaging to the upland game crop.

One of the best and cheapest ways to furnish water to wildlife is to plant mulberries. A surprising number of songbirds, game birds, and mammals eat the berries and feed them to their young. A mulberry tree is also a good protection for orchard fruits and garden truck in danger of bird damage. Tomato-pecking by pheasants, for example, usually represents a desire to drink, not to eat. Mulberry trees bear freely where fertilized by poultry or stock, but in sod or shade the yield of berries is poor. For this reason they are best planted at the farmstead or in fencerows between fields.

Deep water in stock tanks or troughs furnishes drinking water for many birds, but they cannot bathe there unless there is spillage. Wet feathers mean weak flight, hence no bird takes the risk of bathing except in shallow water free from cover which might hide enemies. You can have a bird "bathing beach" on your lawn by building a very shallow concrete pool and filling it every few days. Let the edges "shelve out"; do not build a steep margin.

We seldom think of freshet water on stream bottoms as an asset to wildlife, but if it stands long enough to grow a crop of aquatic animals it constitutes a rich food resource. I have seen young pheasants gorging on snails in a disappearing pond. Gulls, herons, and coons have great tadpole fishing in drying sloughs or ponds.

Many people suppose that wild things in winter eat snow, hence need no water. Yet some do. I have followed deer tracks which made a considerable detour to a snowed-over spring, which had been pawed out by the deer, evidently for drinking.

Some winter behavior in respect to water is so unexpected as to be confusing. I have seen both pheasants and quail during blizzards, wading in unfrozen springs up to their "knees." Were they "warming" their feet in the relatively warm water? Or were they hunting snails to eat? Or were they eating watercress? I am not yet sure.

Bobwhite

JUNE WITHOUT bobwhites whistling in the fencerows is not really June, but only an imitation of it. Yet many a southern Wisconsin farmer, fond of his quail, has himself evicted this bird from his acres without being aware of how or why.

From the time of first settlement to the Civil War, bobwhites were so abundant in the southern counties that farmers trapped and shipped carloads to the big cities. Today, after mild winters, we have a few, and after hard winters, still fewer. Two changes in the land are, I think, responsible for the decline of quail: smaller and fewer weeds due to declining fertility, and smaller and fewer thickets due to elimination of fencerows and pasturing of woodlots. The tall weeds were winter food, and the thickets winter cover.

For weeds we can substitute corn, but for thickets, there is no substitute. We can, however, bolster deficient thickets with grape tangles, brushpiles, and evergreen plantings. Most farms must either build such cover, or go quailless into the future.

Some lean on artificial restocking as a means to more quail, but they lean on a slender reed. Wisconsin is "furthest north" for bobwhite. Any Wisconsin landscape unable to sustain wild quail will certainly fail to sustain pen-raised birds. Pen-raised quail are of southern origin. Wisconsin quail weigh an ounce or two more than southern birds, hence the liberation of southern stock is a positive detriment to the native acre.

Everyone who hears bobwhite knows that he contributes to the satisfactions of farm life, but few know that the Wisconsin bird has contributed to science. It was from intensive studies of quail populations at Prairie du Sac, in Sauk County, that two important principles of wildlife conservation were drawn. One is the principle of carrying capacity: No matter how many quail start the winter on a given area, only a certain number will survive in spring, and that number is the carrying capacity of that area. Carrying capacity can be raised by improving food and cover but not otherwise.

The other is the principle of compensatory mortality: The killing of individual quail by predators does not reduce the surviving population next spring, for if one thing doesn't kill the surplus, something else will, but only down to carrying capacity. Carrying capacity, then, is a threshold of security, below which only catastrophes (like a killing winter) can accomplish a further reduction.

No one knows, as yet, to just what other birds or mammals the two principles apply. No one knows, as yet, just how they operate in quail. They represent the gropings of science toward an understanding of the inner mechanisms which regulate populations. Science knows what chemical elements occur in each star, but not why one species shrinks while another becomes a pest. If more scientists were farmers we might make faster headway on the second problem.

Fifth Column of the Fencerow

Somebody once noticed that bad farmers had big fencerows, and jumped to the conclusion that good farmers should have none at all. Bushy fencerows, it was said, harbor noxious insects and weeds.

It is time to re-examine the soundness of the all-too-simple rule-of-thumb.

Bushy fencerows do harbor noxious insects. They also harbor beneficial insects which prey on the noxious ones, and birds which prey on both. Several hundred kinds of insects, good and bad, and about fifty kinds of birds, mostly beneficial, harbor in fencerows.

The net effect on crops is usually too complex to determine, and never twice the same. In the few instances where the effect is clear, the fencerow is as apt to be an asset as a liability. Thus, pea aphids are less severe near brush because their insect enemies harbor there, but chinch bugs are more severe because they winter in bushy cover. In the majority of instances, the net effect of the fencerow is neutral, and he who likes birds can safely leave fencerow cover for their use.

Bushy fencerows likewise harbor noxious weeds, but they are much less likely to do so than bushless ones, unless the latter be mowed. To understand the reason for this, it is necessary to understand the plant succession.

Just as human pioneers consisted successively of fur traders, settlers, general farmers, and specialized farmers, so plant pioneers consist successively of annuals, grasses, brush, and timber. You can set back this succession with tools, and thus start it over again (as we do each year in plowing), or you can skip to a later stage by planting it artificially (as when we plant a forest), but you cannot change the order of the successive stages on idle ground, for each stage prepares the soil for its successor, and each successor is equipped to displace its predecessor. Thus annuals (such as pigeon grass) are the first to seize naked soil because their seeds are stored in the soil. Perennial herbs, such as aster or goldenrod, displace the annuals because they endure shade in youth, use the annuals as a nurse crop, and being able to winter live roots, they get a running start in the spring. Grasses displace forbs because they come up under them and then choke them by underground runners. Brush displaces grass by overtopping it and shading it out, and brush in turn is overtopped by timber trees.

Most noxious weeds are annuals and hence cannot start on ground already occupied by grass, brush, or timber. The safest way to exclude them from fencerows is to keep the fencerow in solid unbroken sod or brush. Most noxious weeds are intolerant of shade, and hence are excluded by brush.

On the ends of fields where agricultural implements are turned around during cultivation, the wounding of the sod may admit noxious weeds; in such places mowing is the only control. In any case, fewer noxious weeds start in sod than would start on bare ground. In brush noxious weeds cannot start except in openings or on edges.

In my opinion our worst fencerow risk is not weeds, but quack grass, which fights the more desirable grasses with their own weapons, and in distributed fencerows usually wins. The oftener the fencerow is cleaned, the greater the chance of its ultimate invasion by quack. Southern Wisconsin, under the clean fencerow regime, is becoming a lacework of quack, each thread protected by the overhead wires of a fence. Highway engineers, by the repeated grading and filling of roadsides, are unwittingly operating a base from which quack invades the adjoining fields. Like the foreign carp which is quietly displacing the native fishes of our lakes and streams by his superior voracity and reproductive powers, so is this foreign grass quietly displacing the native plants of our roadsides and fencerows by its superior powers of suckering, seeding, and choking. Quack, not insects or weeds, is usually the real fifth-columnist of the fencerow. It is the cleaned fencerow, "softened" by wounding its sod or by removing its bushy shade, which gives quack its initial foothold.

Roadside Prairies

A GOOD TEST of "education" would be to ask a hundred people what is meant by prairie.

Most, I fear, would answer that a prairie is a flat monotonous place good for sixty miles per hour.

A few, I hope, would know.

Tall corn and fat Herefords are prairie symbols. They symbolize the greatest mass effort in human history to extract a rich life from a rich soil.

Pasque flower and blazing star are also prairie symbols. They symbolize the greatest mass effort in evolutionary history to create a rich soil for man to live on. Yet how many farms possess, or cherish, a remnant of them? Just as the barbarians burned the libraries which explained the origins of human culture, so have we plowed under the prairie plants which explain the origins of our prairie empire.

Any prairie farm can have a library of prairie plants, for they are drouth-proof and fire-proof, and are content with any roadside, rocky knoll, or sandy hillside not needed for cow or plow. Unlike books, which divulge their meaning only when you dig for it, the prairie plants yearly repeat their story, in technicolor, from the first pale blooms of pasque in April to the wine-red plumes of bluestem in the fall. All but the blind may read, and gather from the reading new lessons in the meaning of America.

The prairie plants are tough; they ask no quarter of wind or weather, require no pampering with hoe or sprinkler. Nothing can whip them except the overhead shade of trees or sweet clover, the creeping stolons of quack grass, or the continual cropping of cows. Just why the prairie plants stood up under grazing by buffalo and elk, but now succumb to cows, is a mystery. Perhaps the answer is barbed wire, which keeps the cows too long in one place.

Any prairie is a model cooperative commonwealth. Unlike agricultural plants, which "hog" water at random, and devil take the hindmost, each prairie species draws its sustenance from a different subterranean level, so that feast and famine are shared by all species alike. The leguminous members of the community (such as

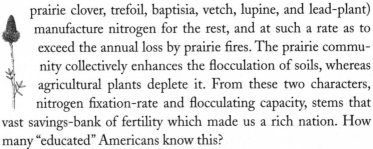

prairie clover, trefoil, baptisia, vetch, lupine, and lead-plant) manufacture nitrogen for the rest, and at such a rate as to exceed the annual loss by prairie fires. The prairie community collectively enhances the flocculation of soils, whereas agricultural plants deplete it. From these two characters, nitrogen fixation-rate and flocculating capacity, stems that vast savings-bank of fertility which made us a rich nation. How many "educated" Americans know this?

The best way to start a library of prairie plants is to find the spot which contains a remnant, and then build up other species around it. Prairie dock and bluestem grass are commonly the last survivors. Most of the prairie species can be grown in the garden from seed and later transplanted to the wild; the University Arboretum has thus established some thirty of them. Planting stock can also be obtained where highway construction is destroying wild remnants. It is wasteful to dig wild stock without expert advice, for the roots of some species go down fifteen feet. It is a sad commentary on our Americanism that the prairie flowers are ignored by commercial seed-dealers and nurseries, and there is no literature on how to grow them. He who learns how is truly a pioneer.

We have thousands of miles of roadside, the outside edges of which are often too steep or rough to mow, already fenced against cows, and kept cleared of brush to prevent snowdrifts. Most of this potential prairie garden is being faithfully stirred up by road-building machinery, after which it goes over, for keeps, to quack grass and sweet clover. Why not let these edges alone and replant them to prairie?

FALL

Wild Foods

Nearly every farming operation offers chances either to conserve or to destroy the wild plants on which game, fur, and feather depend for food. There are many of those wild plants, and each needs its own conditions of soil and light in order to prosper and bear fruit. Each has its own "customers" who like its products.

Winter Fruits. The most dependable yielder is the wild grape. Some vines are male and do not bear. The fall is the time to mark the bearing vines. All they need is full sun and some brush to climb on. Most game and songbirds relish wild grapes, both when fresh and as dried "raisins" in winter.

"Wild" apples are valuable, especially to ruffed grouse, foxes, and deer. The fallen fruit is eaten even after it has frozen. In addition, deer eat the fallen leaves.

All the haws, crabs, viburnums, dogwoods, and sumacs yield valuable winter fruits. All they need is protection from axe, cow, and fire.

Summer Fruits. All game and most songbirds feed berries to their young. The heaviest yielder among the summer berries is the mulberry. If you have none, you can make no better move for wildlife than to plant a few. They bear only in full sun. Mulberries near the garden help draw birds from tame fruits.

The black cherry tree, the choke-cherry bush, the wild plum, and the elder bush are all heavy yielders of summer fruit for wildlife. Protection from fencerow fires and highway crews will usually insure an ample supply.

Fall and Winter Seeds. Almost any fertile soil, when cultivated or otherwise broken up, produces weed seeds valuable as fall and winter food, but the kind of seed depends on the kind of soil.

Thus any marshy spot grows smartweed after cultivation. Smartweed seeds are relished by most game birds, including waterfowl. But marsh soils, if deeply burned, produce only worthless nettles or giant ragweed.

Any upland soil, if fertile, produces foxtail or ragweed after cultivation. Both plants yield abundant seeds which are the staple winter food of quail, pheasants, Hungarian partridge, juncos, and tree sparrows. A supply of seeds can be insured by leaving a strip of oat stubble unplowed till spring, or by omitting the last cultivation on the border of a cornfield.

Peat haymeadows surrounded by brush offer excellent winter cover but usually no food. A heavy crop of false climbing buckwheat can often be secured by lightly burning small spots in the brush. The vines climb the dead brush and hang their seeds out of reach of snow. Broadcast burning for buckwheat, however, is unwise, for the fire may destroy the brush cover, or also it may bite too deep and result in nettles. Spot burning should be rotated to give an annual supply of buckwheat.

Oak woods are sure to yield a small but highly valuable crop of trefoil beans (sometimes called stick-tights) provided the cattle are excluded. Once grazed, however, a woods loses its trefoil for five or ten years. Trefoil beans are the first choice of quail, pheasants, and ruffed grouse for a winter meal. Though carried on a slender stem, the stem is seldom broken down by snows.

Insect Foods. To attract warblers to the farmyard, it is well to have at least one box elder tree. Box elder seems to draw the insects they want, much as clover draws insects for poultry.

Feeding Stations

In feeding birds, as in feeding folks, the first thought is to see that nobody goes hungry.

Experience brings a second thought: to see that feeding does not become too easy for the good of the fed.

The first Wisconsin game bird feeders were hoppers. The birds stood up to a tray and gorged.

It is better for a bird to gorge than to starve, and hoppers are still recommended where the operator is unable to visit the station frequently. A hopper set under a roofed shelter (so as to keep ice out of the grain) is nearly automatic, and a single filling often lasts for weeks.

But where the operator can tend the station frequently (say twice a week) a hopper is far inferior to a straw bed.

A straw bed is simplicity itself. Build a roofed shelter facing east or south, put straw or chaff on the floor, and throw shelled corn into it. Let the birds scratch and earn their keep. To notify the birds that there is grain in the straw, lay a few cobs on it, or scatter "bait" grain on bare ground nearby.

Do not use marsh hay for the scratching bed; it's too heavy. Quail can't move heavy materials.

European farmers discarded hoppers long ago, and feed mostly in straw beds. The exercise is good for the birds; there is much less wastage by squirrels and rabbits. The only limitation on the straw-bed method is that fresh grain must be added every few days.

If you have saved the chaff from under the corn shredder, it is superior to straw, especially for winter songbirds. The foxtail and ragweed seeds contained in such chaff are attractive to pheasants, quail, and Hungarians as well. Beware of moldy chaff; it must usually be dried to keep well.

Don't build your shelter too tight; it should have plenty of escape holes, else the birds may refuse to enter it.

The consumption of grain from hoppers runs about as follows: two pounds per pheasant per week; one pound per prairie chicken per week; three-quarter pound per Hungarian per week; one-half pound per quail per week. Rabbits and squirrels eat two pounds or more per week, and squirrels may waste an almost unlimited amount by taking out the "hearts" of the corn.

Soy-beans, buckwheat, wheat, rye, barley, and sorghum are all acceptable substitutes for corn. Cracked acorns are also good.

One-half of 1 per cent of the corn raised on an average southern Wisconsin farm will winter a good stand of wildlife. Does wildlife add one-half of 1 per cent to the satisfaction of rural living?

Even if pheasants are not "planted" on the farm, there is profit in encouraging them and other birds to stay. Feed is plentiful this season and will scarcely be missed from stores for livestock. It does not require any great outlay of time or labor or feed to feed the game birds in this way this winter.

Look for Bird Bands

DEAD BIRDS which have been shot, or found killed on high-ways, sometimes bear leg bands of aluminum or colored celluloid. Each band bears a number, and usually a return address.

Such birds have been banded for a definite purpose. Wild birds are trapped, banded, and released in the hope of tracing their migration routes. Birds raised in captivity are banded upon release in the hope of tracing their survival and movements. Banding also yields information on how long birds live, on their rate of mortality, and on the composition of populations as to sex and age.

A banded bird invariably represents a lot of work done by some scientist or conservation officer, but that work comes to nothing unless the finder reports where, when, and, if possible, how the banded bird was killed. Banding depends for its success on the cooperation and good will of the public.

This is a plea for your cooperation and good will in looking for bird bands, and in reporting them to the address given on the band. Most Wisconsin bands are marked for return either to the Conservation Department, Madison, or to the United States Biological Survey, Washington, D.C. If in doubt, send the band to the Conservation Department. You will receive a return report telling you where, when, and by whom your bird was banded.

If you must keep the band for a watch-fob, do so, but send in the number just the same, with a letter telling where and when the bird was killed.

Here are some examples of valuable facts gleaned from banding reports.

A farmer near Ladysmith found a banded prairie chicken dead on the road in summer. It had been banded two years before at La Crosse, in winter. This one bird answered a question which had been in dispute for years, do Wisconsin chickens ever migrate? This bird, a hen, had migrated a hundred miles.

The farmers of the Riley Game Cooperative had been planting pheasants for years, but by winter the pheasants mostly disappeared. It was supposed that they moved eastward toward the Sugar Creek marshes. By banding the birds and then noting where the hunters killed them, it was found that the main movement was westward, not eastward.

A farmer in Sauk County trapped and banded all the chickadees which came to his feeding station. After a dozen had been marked, no more unmarked birds appeared in the trap. This gave a reliable census of the local population. Next summer several marked chickadees were seen nesting nearby. This showed that part, at least, of the summer chickadees were yearlong residents. Next winter about half the original dozen reappeared at the feeder. This showed that during the year half of the population either moved or died. All these were new and hitherto unknown facts about chickadees.

Banding, in short, enables us to see things in nature which are otherwise invisible, even to the most skilled woodsman. To promptly report all bands is one of the ways in which we can all help build a sounder knowledge of wildlife, and a sounder conservation policy.

The Hawk and Owl Question

THERE IS an old saying that the only good hawk or owl is a dead one, but the more this attitude is examined in the light of science, the less truth is found in it.

All hawks and owls eat what they can catch, but few can catch healthy game or grown poultry. Of the six common kinds of hawks and five common owls, only four ordinarily catch game or chickens. The rest do not "abstain"; they are simply unable to connect. All kinds live principally on mice, gophers, rabbits, snakes, frogs, small birds, and other abundant food.

It has become customary to say that horned owls, Cooper's hawks, and sharpshin hawks are bad, but that the rest are beneficial. This implies that the countryside would be better off without these three "bad" kinds. It is certain that the countryside would be worse off without them, for the same skill and speed which enable them to do damage make them especially valuable as checks on small animals. Much the best rule of action is to shoot any hawk or owl actually caught doing damage, but no other.

Hawks and owls, during the nesting season, are intense devotees of "private property"; they tolerate no trespass by others of their kind. It is a sad thing to see a farmer shoot on suspicion a horned owl that has been doing him no harm, thereby opening the woodlot to another that may promptly invade his poultry pen. Individual hawks and owls differ in their habits, and a farm inhabited by individuals which do not have the poultry habit is automatically protected, during much of the year, against those that do.

The old belief that abundant game can be had only by shooting off the hawks and owls has been thoroughly discredited. They eat game, to be sure, but they are ordinarily able to catch only the excess above what the cover can carry. Such excess game is constantly built up by natural increase. Moderate preying on game has no more effect on next year's game population than moderate shooting does. On the other hand the total removal of hawks and owls exposes game to keener competition from rodents which eat, in general, the same foods. This university has four experimental areas where the stand of game has been built up to high levels

solely by improving food and cover. No hawk or owl has ever been shot on them, but game builds up year after year to as high a level as the food and cover allow. These areas are living proof that to remove hawks and owls for the protection of wild game is unnecessary.

Feed Early

Notions of how to feed wildlife in winter are changing rapidly. One of the notions now under suspicion is the permanent feeding shelter. All feeding stations which stay in exactly the same location year after year may harbor parasites and diseases, and thus do more harm than good. It is best either to use a "natural" feeder, like corn-shocks, or a portable one, which can be moved to fresh ground each year.

Among the good portable feeders are the spiked-pole rack and the woven wire basket. The first consists of a pole through which long spikes are driven, and corn ears impaled on the spikes. The pole may be set up against a tree or fence, or raised on forked poles near cover. The basket consists of a cylindrical roll of hog-wire filled with ear corn. Neither of these devices requires a shelter, but for feeding quail they should be located near dense cover, such as brushpiles, grape tangles, or evergreen thickets.

The more experience accumulates, the clearer it is that feeding should begin early, should offer only moderate amounts of grain, and should be kept up all winter with religious regularity. The farmer can do such feeding because he lives nearby. Those who can tend their feeders only on weekends have a harder problem.

It is now clear that wintering coveys of game birds are led by "old-timers" who remember where the feed was last winter. Your reputation for dependability is, therefore, almost as important in handling birds as in handling bankers. It took me six years to re-establish the reputation of my farm as a place for birds to winter. Quail and pheasants nest there regularly, but my predecessor was more interested in bootleg than in birds, hence the birds "knew" my place was no place to stay after snow flies. Autumn after autumn, by the time I got my feed out, they were gone. Finally, by leaving a sweetcorn patch unpicked, I got them started feeding in September, and they became aware of the adjacent field corn grown for their benefit.

This theory that old-timers set the social stage among birds is no mere supposition. Banding studies show that a few old birds are present in most flocks, and these are believed to teach the rest not only where to feed, but where to seek cover, and how to escape enemies. Many investigators have noticed how game birds will stick to what was a good covert, even after it has been cut off or grazed out. Tradition, in short, remains an important social force, revolutions to the contrary notwithstanding.

Farm Fur Crops

Farm fur has always meant pin-money for the farm boy. It has also meant a kind of education not available in schools, but very important in the building of our peculiar national character. The boy-trapper is one of the principal components in what I call split-rail Americanism.

Farm fur of the more valuable species has been declining in most localities, and many farms are already in the furless class. On such farms the boy must do his Daniel-Booneing in books.

On farms which still have fur animals, what can be done to cultivate and perpetuate the crop?

Marsh drainage, woodlot grazing, and stream-straightening tend, as a rule, to reduce the kinds and numbers of fur animals. Sometimes the fur-income from a marsh exceeds the potential crop-income, especially if it be good habitat for muskrats. What makes a marsh productive of rats?

First, trapping must not be so heavy as to remove the breeding stock. Rats breed with great rapidity, and a thin stock breeds more rapidly than an overcrowded one, but a ratless marsh must await the influx of new stock before it can produce at all.

Second, rats should be protected, if possible, from sudden draw-downs of water-level in winter. Rats cannot feed when the bottom freezes. Falling ice entraps rats in their houses, where they freeze and starve, or fall prey to minks.

Third, rats should have either bank burrows, or vegetation for house-building, or both. Houses are erected in late October, and

require cattails, or other bulky abundant vegetation, for their construction. Some, but not all, house materials are also good food materials.

Fourth, rats should have winter food. In Wisconsin cattail, sedges, bugleweed, sweet flag, and "three-square" rush are sample winter food plants of high value. Aquatic animals, especially clams, are eaten in winter, but are seldom abundant enough to be standbys. Corn, vegetables, and dried land-vegetation are taken in a pinch, but all these dry-land foods entail exposure, and exposure means losses. Summer foods are of large variety, and seldom present a problem.

Minks, unless overtrapped, usually occur wherever muskrats do. They prey heavily on rats rendered defenseless by low water or other exigency. In normal habitat good stands of mink and muskrat inhabit the same water, just as good stands of upland game cohabit with numerous hawks, owls, and foxes. (If anybody doubts this, take a look at the University of Wisconsin Arboretum, where nothing is "controlled," and nearly everything is abundant.)

Raccoons, nowadays, are limited mainly to the vicinity of rock ledges, drain-tiles, and other impregnable dens. This fact tells its own story: We have chopped and dug out most of the natural coon dens, and then hounded and trapped to death the defenseless residue of our breeding stock. On my own farm there are always coon tracks in fall, seldom in spring. The den trees opened or felled by self-invited trappers and hunters tell why. But for half-uprooted trees tipped over by floods or root-diseases, my coons would disappear entirely, despite my efforts or root-diseases.

Opossums and gray foxes do not have to be worried over. Red foxes are another matter: They seem to shrink with the woodlots and fencerows. Why worry about foxes at all? Count the mice in the stomach of your next fox, and you'll know the answer. Better still, find where some fox has vomited a half-a-hatful of dead mice from his overloaded stomach. Besides, fox-hunting (not with pink coats, but American brand) is one of the outstanding split-rail sports. When the farm boy can no longer see a fox-track in the snow, it will take a lot to replace the loss. When fox-hunting, native style, is over with, a certain glory will have vanished from the earth.

Farming in Color

At his best man has a hard job improving on nature.

On Highway 12, a few miles south of Baraboo, a ledge of pink quartzite forced the engineers to leave a hairpin turn in the road. Backed up against this lodge, like a soldier making a last-ditch stand, is an old sugar maple. He is nothing much to look at during spring, summer, or winter. Pretty scrubby to start with, he is plastered from root to branch with the scrubbiest of Americana: the advertising sign.

Some highway engineer, for good measure, dynamited the ledge and tore off many of his twigs and limbs; his foliage pattern reminds one of an old crow with his wingfeathers all but shot out. As a last ignominy, somebody painted an advertisement on the rock behind the tree.

But come October, and this old veteran shines with a glory that all these banalities cannot efface. He is crowned with a crimson halo that fades only with the onset of heavy frosts, and he will wear his October crown each year until the engineers and advertisers have made an end of him.

There is nothing particularly noteworthy about a well-colored maple; Wisconsin has more of them than any region on earth, barring possibly Vermont. The noteworthy thing is the blindness of those who live with this old veteran but have never noticed him.

Many farms, as well as highways, have lost their well-colored maples. Not all farmers are blind. Some may wish to restore October color to their farms. How does one go about it?

A sugar maple takes many decades to grow. A quicker-growing substitute is the red maple, or swamp maple. The red maple lacks the subtlety and variety of the sugar maple, but produces a good uniform crimson wherever he will grow at all. Acid soils are best. The edge of the woods, especially the edge of a pine plantation, is the best background. Red maple stock is available at the State Nursery as "wildlife cover," and at many private nurseries.

Sugar maples color best on soils which grew them naturally, i.e., on soils which are fairly heavy, and which are too acid to grow clover without lime. Sugar maple is a valuable timber tree, and good for underplanting in woods too dense to grow young oaks. The tree is liable to sunscald in hot windy locations.

Wisconsin has many shrubs good for color-farming. One of the most striking is the native wahoo, whose berries hang like cerise lanterns in October sun. Aspen often turns a fine clear lemon-yellow which is effective, especially against dark pines. Aspen, strangely enough, often fails to grow after being transplanted. Such a scrubby aspen, if chopped off and forced to resprout, may make a straight white bole.

Foreign trees and shrubs, almost without exception, give no color in fall. The leaves simply freeze off because the species is not quite adapted to our climate. Apple and lilac are examples.

One of the irrefutable indictments of Wisconsin's "taste in natural objects" is the fact that our capitol square is planted with European maples. I have yet to hear a 4th of July orator who has noticed this.

Smartweed Sanctuaries

Many a farmer feels a friendly interest in ducks, but lacks a pond or other permanent water on which to offer them hospitality. To such I recommend the smartweed sanctuary as a possible substitute.

To start such a sanctuary you need two things: a pothole or other low spot which will hold temporary water in spring and fall, and a source of water to fill it.

The water is the rub. Sometimes a creek or spring can be diverted into the pothole. In wet years a dry channel may flow often enough to be led into it. Western farmers have learned to lead water around by the nose, but in this region we seldom consider the possibilities of water diversion.

If you can contrive your water supply, the rest is easy. Any low spot with fairly good soil will grow smartweed if the sod is broken up each fall or spring by plow, harrow, or rooting swine. Flood it in March when the spring flight begins. The ground should be moist, but not submerged, in June, July, and August. Flood it again in September when the first teal come down, and keep all guns away. You may soon have a duck-show worth getting up early to see. Keep this up year after year, and your duck-show grows better as more and more ducks learn where they may feed in safety.

Smartweed need not be flooded deeply to draw ducks. River ducks prefer water between an inch and a foot in depth. Diving ducks prefer deeper submersion. Both classes of ducks eat smartweed seeds with relish.

Smartweed seeds, as every observant farmer knows, are stored in the soil of all low spots. Break the sod, and the smartweeds flourish like a green bay tree. There are many kinds of smartweeds, but all bear a heavy crop of flat, blackish, shiny seeds which most ducks will fly miles to guzzle, provided they aren't met on their arrival with an anti-aircraft barrage. Even sedge marsh without a single visible smartweed plant will sometimes provide enough stored smartweed seed to draw and feed a few ducks.

Sometimes low ground, when broken, comes up to Spanish needle or "pitchfork" weed *(Bidens)* instead of smartweed. This weed is much less desirable as duck food. No one yet knows just how to regulate the time and manner of breaking so as to get smartweed and not pitchforks. Farmers who can learn how to turn this trick will make a valuable contribution to wildlife management.

A bed of smartweed, flooded or unflooded, draws the pheasant as a cookie-jar draws the schoolboy. When offered flooded smartweed, the pheasant acts as if he had grown webs between his toes; you may flush him from water a foot deep, but his plumage is dry. I think the answer is that he walks on the masses of floating stems, as a rail or gallinule does. When one examines the crop of such a "sea-going" pheasant, one finds it full of smartweed seeds, or else it is full of slugs, snails, or other water-animals which multiply promptly on flooded ground.

Should you want to learn the various kinds of smartweed, and what kinds are eaten by what ducks, I suggest you send for U.S.D.A. Technical Bulletin 634, "Food of Game Ducks in the United States and Canada."

From Little Acorns

"GREAT OAKS from little acorns grow" if—the acorns have no worms in them, if—they get planted in the right soil at the right time, if—a gopher or squirrel doesn't dig them up, and if—the seedling does not get choked by grass sod, or nipped by a cow, or girdled by a rabbit, or shaded out by brush or trees, or burned by fires.

Men, as well as oaks, have to run a gauntlet of "ifs." Perhaps that is why men of understanding cherish their trees.

Oaks have a precarious and recently gained foothold on our southern Wisconsin hills. Early travelers, with one accord, describe as nearly treeless many a locality now plentifully sprinkled with woodlots. One of the clearest of these early descriptions is the journal of Captain Jonathan Carver, who crossed the state in 1766, a decade before the Revolutionary War.

Arriving at Green Bay, and ascending the Fox River toward lake Winnebago and Portage, he says: "The country is fertile, and in no part very woody." The "carrying place" between the Fox and the Wisconsin (now Portage) he describes as "a plain with some few oaks and pine trees growing thereon."

Descending the Wisconsin to Prairie du Sac, Carver "took a view of some mountains that lie about fifteen miles to the southward, and abound in lead ore. I ascended one of the highest of these (Blue Mound?) and had an extensive view of the country. For many miles nothing was to be seen but lesser mountains, which appeared at a distance like hay cocks, they being free from trees. Only a few groves of hickory and stunted oaks covered some of the valleys." Carver was looking at the region from Verona to Ridgeway, and from Black Earth to New Glarus. It is now well wooded.

One must conclude from this that most of our present woodlands in the southern counties sprang up since Carver's day; probably when the early plowings had broken the free sweep of prairie fires. Fire, in Carver's day, was the big "if" for little acorns.

One must also conclude that an oak more than a century old is really an historical monument, and should hardly be cut for ordinary bread-and-butter purposes.

That there was a widespread encroachment of oaks on prairie, especially during the 1850's, is attested by the transactions of the Wisconsin Agricultural Society, which contain lengthy debates about where the new oak seedlings all came from, and why most of the new seedlings were red oaks, whereas most of the seed trees were bur oaks. Some of the debaters opined the wild pigeons had brought the acorns in their crops, and seeing plenty of wheat to eat, had regurgitated them. Others argued that spontaneous generation must be accountable. (Education has made progress!)

The farmer can often verify for himself whether his woodlot sprang up after settlement, or whether woods existed there prior to settlement. He can do this by counting the annual rings on freshly cut stumps, and by sawing off the butt-log from natural windfalls. Most farm woodlots in the regions which were once prairie now show a preponderance of 80-year-old oaks, plus a sprinkling of old veterans, often 150 to 200 years old. These veterans are commonly branchy, showing that they grew up on open prairie. The stumps often show early scars of prairie fires. These are the trees that Carver saw.

On the other hand the 80-year-old trees are cleaner boled, showing that they grew up in a denser stand. These are the trees that came in when the prairie fires ceased.

It is clear, then, that southern Wisconsin woodlots spread out during the 1800's. It is equally clear that they are now shrinking. The main reason is pasturing. Heavy pasturing presents a hopeless "if" in the gauntlet which oak seedlings must run. As old oaks are cut from pastured woods or die off of drouth and disease, they are not replaced by young ones. In the long run this means either an almost treeless landscape, or one decorated only with box elder, Chinese elm, locust, and other weed trees.

It has recently come to light that oak woods which have been fully protected reproduce to oaks only when pretty heavily opened up. Light scattered cuttings bring not oak seedlings, but maple, elm, basswood, and other shade-enduring hardwoods. This indicates that oak-hickory woods is a temporary type, which, if let alone, changes over gradually to mixed hardwoods of the kind found in the eastern and northern counties.

Woodlot Wildlife and Plant Disease

EVERY FARM woodlot, in addition to yielding fuel and posts, should yield its owner a liberal education. This crop of wisdom never fails, but it is not always harvested.

I here record one of the many lessons I have learned in my own woodlot.

Soon after I bought the land a decade ago, I realized that I had bought almost as many tree diseases as I had trees. The place is riddled by all the ailments wood is heir to. I began to wish that Noah, when he loaded up the Ark, had left the tree diseases behind.

My woods is headquarters for a family of coons; few of my neighbors have any. One snowy Sunday I learned why. The fresh track of a coon-hunter and his hound led up to a half-uprooted maple, under which one of my coons had taken refuge. The frozen snarl of roots and earth was too rocky to chop and too tough to dig; the holes under the roots too numerous to smoke out. The hunter had quit coonless because a fungus disease had weakened the roots of the maple. The tree, half-tipped-over by a storm, offers an impregnable fortress for coondom. Without this "bombproof" shelter, my seed stock of coons would be cleaned out by hunters each year.

My woods houses a dozen ruffed grouse, but my pines are all too young to make cover, hence in deep snow my grouse shift to my neighbor's woods, where there are several pine thickets. However, I always retain as many grouse as I have oaks wind-thrown by summer storms. These summer windfalls keep their dried leaves, and during snows each such windfall harbors a grouse. The droppings show that the grouse roost, feed, and loaf, for the duration of the storm, within the narrow confines of their leafy camouflage, safe from wind, owl, fox, and hunter. The cured oak leaves not only serve as cover, but for some curious reason are relished as food by the grouse.

These oak windfalls are, of course, diseased trees. Without disease, few oaks would break off, and hence few grouse would have down tops to hide in.

During October, my grouse often stuff themselves with oak galls, a diseased growth caused by the sting of a gall-wasp.

Each year the wild bees load up one of my hollow oaks with combs, and each year trespassing honey-hunters harvest the honey before I do. This is partly because they are more skillful than I am in "lining up" the bee trees, and partly because they use nets, and hence are able to work before the bees become dormant in fall.

But for heart-rots, there would be no hollow oaks to furnish wild bees with hives.

During high years of the cycle, there is a plague of rabbits in my woods. They eat the bark and twigs off of almost every kind of tree or bush I am trying to encourage, and ignore almost every kind I

would like to have less of. (When the hunter plants himself a grove of pines or an orchard, the rabbit somehow ceases to be a game animal, and becomes a pest instead.)

The rabbit, despite the catholicity of his appetite, is an epicure in some respects. He always prefers a hand-planted pine, maple, apple, or wahoo to a wild one. He also insists that certain salads be pre-conditioned before he deigns to eat them. Thus he spurns red dogwood until it is attacked by oyster-shell scale, after which the bark becomes a delicacy, to be eagerly devoured by all the rabbits in the neighborhood.

A flock of a dozen chickadees spends the year in my woodlot. In winter, when we are harvesting our fuelwood, the ring of the axe is dinner-gong for the chickadee tribe. They hang in the offing waiting for the tree to fall, offering pert commentary on the slowness of our labor. When the tree at last is down, and the wedges begin to open up its contents, the chickadees draw up their white napkins and fall to. Every slab of dead bark is, to them, a treasury of eggs, larvae, and cocoons. For them every split of grub- or ant-tunneled heartwood bulges with milk and honey. We often stand a fresh split against a nearby tree just to see the greedy chicks mop up the ant-eggs. It lightens our labor to know that they, as well as we, derive aid and comfort from the fragrant riches of newly split oak.

But for tree diseases and insect pests, there would likely be no food in these trees, and hence no chickadees to add cheer to winter woods.

Many other kinds of wildlife depend on tree diseases. My pileated woodpeckers chisel living pines, to extract fat grubs from the diseased heartwood. My barred owls find surcease from crows and jays in the hollow heart of an old basswood; but for this diseased tree their sundown serenade would probably be silenced. My wood ducks doubtless nest in hollow trees; at least they do not nest in the artificial houses I have put up, and every June brings its brood of downy ducklings to my woodland slough. All squirrels depend, for permanent dens, on a delicately balanced equilibrium between a rotting cavity and the scar-tissue with which the tree attempts to close the wound. The squirrels referee the contest by gnawing out the scar-tissue when it begins unduly to shrink the amplitude of their front door.

The real jewel of my disease-ridden woodlot is the prothonotary warbler. He nests in an old woodpecker hole, or other small cavity, in a dead snag overhanging water. The flash of his gold-and-blue plumage amid the dank decay of the June woods is in itself proof that dead trees are transmuted into living animals, and vice versa.

When you doubt the wisdom of this arrangement, take a look at the prothonotary.

PART III

Conservation and Land Health

The Farmer as a Conservationist

In this masterpiece, published in 1939 in American Forests, *Leopold describes conservation as a "harmony between men and land" and a quest for "wholeness in the farm landscape." He presents as potent impediments to conservation the mentality of judging land use by a single measure, such as crop yields, and the "regimentation of the human mind" brought on by "our self-imposed doctrine of ruthless utilitarianism." Leopold here appeals directly and powerfully to the hearts and souls of farmers, calling on them to rekindle their curiosity about nature and to recapture the drama of farm life. He ends the piece by stepping into the future and describing, in alluring, lyrical terms, what "the landscape of a corn-belt farm" might one day look like in the hands of a caring, skillful farmer, a farmer who remembers with disdain the old days when "everybody worried about getting his share; nobody worried about doing his bit."*

CONSERVATION means harmony between men and land.

When land does well for its owner, and the owner does well by his land; when both end up better by reason of their partnership, we have conservation. When one or the other grows poorer, we do not.

Few acres in North America have escaped impoverishment through human use. If someone were to map the continent for gains and losses in soil fertility, waterflow, flora, and fauna, it would be difficult to find spots where less than three of these four basic resources have retrograded; easy to find spots where all four are poorer than when we took them over from the Indians.

As for the owners, it would be a fair assertion to say that land depletion has broken as many as it has enriched.

It is customary to fudge the record by regarding the depletion of flora and fauna as inevitable, and hence leaving them out of the account. The fertile productive farm is regarded as a success, even though it has lost most of its native plants and animals. Conservation protests such a biased accounting. It was necessary, to be sure, to eliminate a few species, and to change radically the distribution of many. But it remains a fact that the average American township has lost a score of plants and animals through indifference for every one it has lost through necessity.

What is the nature of the process by which men destroy land? What kind of events made it possible for that much-quoted old-timer to say: "You can't tell me about farming; I've worn out three farms already and this is my fourth"?

Most thinkers have pictured a process of gradual exhaustion. Land, they say, is like a bank account: If you draw more than the interest, the principal dwindles. When Van Hise said "Conservation is wise use," he meant, I think, restrained use.

Certainly conservation means restraint, but there is something else that needs to be said. It seems to me that many land resources, when they are used, get out of order and disappear or deteriorate before anyone has a chance to exhaust them.

Look, for example, at the eroding farms of the cornbelt. When our grandfathers first broke this land, did it melt away with every rain that happened to fall on a thawed frost-pan? Or in a furrow not exactly on contour? It did not; the newly broken soil was tough, resistant, elastic to strain. Soil treatments which were safe in 1840 would be suicidal in 1940. Fertility in 1840 did not go down river faster than up into crops. Something has got out of order. We might almost say that the soil bank is tottering, and this is more important than whether we have overdrawn or underdrawn our interest.

Look at the northern forests: Did we build barns out of all the pineries which once covered the lake states? No. As soon as we had opened some big slashings we made a path for fires to invade the woods. Fires cut off growth and reproduction. They outran the lumberman and they mopped up behind him, destroying not only the timber but also the soil and the seed. If we could have kept the soil and the seed, we should be harvesting a new crop of pines now, regardless of whether the virgin crop was cut too fast or too slow. The real damage was not so much the overcutting, it was the run on the soil-timber bank.

A still clearer example is found in farm woodlots. By pasturing their woodlots, and thus preventing all new growth, cornbelt farmers are gradually eliminating woods from the farm landscape. The wildflowers and wildlife are of course lost long before the woodlot itself disappears. Overdrawing the interest from the woodlot bank is perhaps serious, but it is a bagatelle compared with destroying the capacity of the woodlot to yield interest. Here again we see awkward use, rather than overuse, disordering the resource.

In wildlife the losses from the disordering of natural mechanisms have, I suspect, far exceeded the losses from exhaustion. Consider the thing we call "the cycle," which deprives the northern states of all kinds of grouse and rabbits about seven years out of every ten. Were grouse and rabbits always and everywhere cyclic? I used to think so, but I now doubt it. I suspect that cycles are a disorder of animal populations, in some way spread by awkward landuse. We don't know how, because we do not yet know what a cycle is. In the far north cycles are probably natural and inherent, for we find them in the untouched wilderness, but down here I suspect they are not inherent. I suspect they are spreading, both in geographic sweep and in number of species affected.

Consider the growing dependence of fishing waters on artificial restocking. A big part of this loss of toughness inheres in the disordering of waters by erosion and pollution. Hundreds of southerly trout streams which once produced natural brook trout are stepping down the ladder of productivity to artificial brown trout, and finally to carp. As the fish resource dwindles, the flood and erosion losses grow. Both are expressions of a single deterioration. Both are not so much the exhaustion of a resource as the sickening of a resource.

Consider deer. Here we have no exhaustion; perhaps there are too many deer. But every woodsman knows that deer in many places are exterminating the plants on which they depend for winter food. Some of these, such as white cedar, are important forest trees. Deer did not always destroy their range. Something is out of kilter. Perhaps it was a mistake to clean out the wolves; perhaps natural enemies acted as a kind of thermostat to close the "draft" on the deer supply. I know of deer herds in Mexico which never get out of kilter with their range; there are wolves and cougars there, and always plenty of deer but never too many. There is substantial balance between those deer and their range, just as there was substantial balance between the buffalo and the prairie.

Conservation, then, is keeping the resource in working order, as well as preventing overuse. Resources may get out of order before they are exhausted, sometimes while they are still abundant. Conservation, therefore, is a positive exercise of skill and insight, not merely a negative exercise of abstinence or caution.

What is meant by skill and insight?

This is the age of engineers. For proof of this I look not so much to Boulder Dams or China Clippers as to the farmer boy tending his tractor or building his own radio. In a surprising number of men there burns a curiosity about machines and loving care in their construction, maintenance, and use. This bent for mechanisms, even though clothed in greasy overalls, is often the pure fire of intellect. It is the earmark of our times.

Everyone knows this, but what few realize is that an equal bent for the mechanisms of nature is a possible earmark of some future generation.

No one dreamed, a hundred years ago, that metal, air, petroleum, and electricity could coordinate as an engine. Few realize today that soil, water, plants, and animals are an engine, subject, like any other, to derangement. Our present skill in the care of mechanical engines did not arise from fear lest they fail to do their work. Rather was it born of curiosity and pride of understanding. Prudence never kindled a fire in the human mind; I have no hope for conservation born of fear. The 4-H boy who becomes curious about why red pines need more acid than white is closer to conservation than he who writes a prize essay on the dangers of timber famine.

This necessity for skill, for a lively and vital curiosity about the workings of the biological engine, can teach us something about the probable success of farm conservation policies. We seem to be trying two policies, education and subsidy. The compulsory teaching of conservation in schools, the 4-H conservation projects, and school forests are examples of education. The woodlot tax law, state game and tree nurseries, the crop-control program, and the soil conservation program are examples of subsidy.

I offer this opinion: These public aids to better private land-use will accomplish their purpose only as the farmer matches them with this thing which I have called skill. Only he who has planted a pine grove with his own hands, or built a terrace, or tried to raise a better crop of birds can appreciate how easy it is to fail; how futile it is passively to follow a recipe without understanding the mechanisms behind it. Subsidies and propaganda may evoke the farmer's acquiescence, but only enthusiasm and affection will evoke his skill. It takes something more than a little "bait" to succeed in conservation. Can our schools, by teaching, create this something? I hope so, but I doubt it, unless the child brings also something he gets at home. That is to say, the vicarious teaching of conservation is just one more kind of intellectual orphanage; a stop-gap at best.

Thus we have traversed a circle. We want this new thing, we have asked the schools and the government to help us catch it, but we have tracked it back to its den under the farmer's doorstep.

I feel sure that there is truth in these conclusions about the human qualities requisite to better land-use. I am less sure about many puzzling questions of conservation economics.

Can a farmer afford to devote land to woods, marsh, pond, windbreaks? These are semi-economic land-uses—that is, they have utility but they also yield non-economic benefits.

Can a farmer afford to devote land to fencerows for the birds, to snag-trees for the coons and flying squirrels? Here the utility shrinks to what the chemist calls "a trace."

Can a farmer afford to devote land to fencerows for a patch of ladyslippers, a remnant of prairie, or just scenery? Here the utility shrinks to zero.

Yet conservation is any or all of these things.

Many labored arguments are in print proving that conservation pays economic dividends. I can add nothing to these arguments. It seems to me, though, that something has gone unsaid. It seems to me that the pattern of the rural landscape, like the configuration of our own bodies, has in it (or should have in it) a certain wholeness. No one censures a man who loses his leg in an accident, or who was born with only four fingers, but we should look askance at a man who amputated a natural part on the grounds that some other is more profitable. The comparison is exaggerated; we had to amputate many marshes, ponds, and woods to make the land habitable, but to remove any natural feature from representation in the rural landscape seems to me a defacement which the calm verdict of history will not approve, either as good conservation, good taste, or good farming.

Consider a single natural feature: the farm pond. Our godfather the Ice-king, who was in on the christening of Wisconsin, dug hundreds of them for us. We have drained ninety and nine. If you don't believe it, look on the original surveyor's plot of your township; in 1840 he probably mapped water in dozens of spots where in 1940 you may be praying for rain. I have an undrained pond on my farm. You should see the farm families flock to it of a Sunday, everybody from old grandfather to the new pup, each bent on the particular aquatic sport, from water lilies to bluegills, suited to his (or her) age and waistline. Many of these farm families once had ponds of their own. If some drainage promoter had not sold them tiles, or a share in a steam shovel, or some other dream of sudden affluence, many of them would still have their own water lilies, their own bluegills, their own swimming hole, their own redwings to hover over a buttonbush and proclaim the spring.

If this were Germany, or Denmark, with many people and little land, it might be idle to dream about land-use luxuries for every farm family that needs them. But we have excess plowland; our conviction of this is so unanimous that we spend a billion out of the public chest to retire the surplus from cultivation. In the face of such an excess, can any reasonable man claim that economics prevents us from getting a life, as well as a livelihood, from our acres?

Sometimes I think that ideas, like men, can become dictators. We Americans have so far escaped regimentation by our rulers, but have we escaped regimentation by our own ideas? I doubt if there exists today a more complete regimentation of the human mind than that accomplished by our self-imposed doctrine of ruthless utilitarianism. The saving grace of democracy is that we fastened this yoke on our own necks, and we can cast it off when we want to, without severing the neck. Conservation is perhaps one of the many squirmings which foreshadow this act of self-liberation.

The principle of wholeness in the farm landscape involves, I think, something more than indulgence in land-use luxuries. Try to send your mind up in an airplane; try to see the *trend* of our tinkerings with fields and forests, waters and soils. We have gone in for governmental conservation on a huge scale. Government is slowly but surely pushing the cutovers back into forest; the peat and sand districts back into marsh and scrub. This, I think, is as it should be.

But the cow in the woodlot, ably assisted by the axe, the depression, the June beetle, and the drouth, is just as surely making southern Wisconsin a treeless agricultural steppe. There was a time when the cessation of prairie fires added trees to southern Wisconsin faster than the settlers subtracted them. That time is now past. In another generation many southern counties will look, as far as trees are concerned, like the Ukraine, or the Canadian wheatlands. A similar tendency to create *monotypes,* to block up huge regions to a single land-use, is visible in many other states. It is the result of delegating conservation to government. Government cannot own and operate small parcels of land, and it cannot own and operate good land at all.

Stated in acres or in board feet, the crowding of all the timber into one place may be a forestry program, but is it conservation? How shall we use forests to protect vulnerable hillsides and river-banks from erosion when the bulk of the timber is up north on the sands where there is no erosion? To shelter wildlife when all the food is in one county and all the cover in another? To break the wind when the forest country has no wind, the farm country nothing but wind? For recreation when it takes a week, rather than an hour, to get under a pine tree? Doesn't conservation imply a certain interspersion of land-uses, a certain pepper-and-salt pattern in the warp and woof of the land-use fabric? If so, can government alone do the weaving? I think not.

It is the individual farmer who must weave the greater part of the rug on which America stands. Shall he weave into it only the sober yarns which warm the feet, or also some of the colors which warm the eye and the heart? Granted that there may be a question which returns him the most profit as an individual, can there be *any* question which is best for his community? This raises the question: Is the individual farmer capable of dedicating private land to uses which profit the community, even though they may not so clearly profit him? We may be overhasty in assuming that he is not.

I am thinking, for example, of the windbreaks, the evergreen snowfences, hundreds of which are peeping up this winter out of the drifted snows of the sandy counties. Part of these plantings are subsidized by highway funds, but in many others the only subsidy is the nursery stock. Here then is a dedication of private land to a community purpose, a private labor for a public gain. These windbreaks do little good until many landowners install them; much good after they dot the whole countryside. But this "much good" is an undivided surplus, payable not in dollars, but rather in fertility, peace, comfort, in the sense of something alive and growing. It pleases me that farmers should do this new thing. It foreshadows conservation. It may be remarked, in passing, that this planting of windbreaks is a direct reversal of the attitude which uprooted the hedges, and thus the wildlife, from the entire cornbelt. Both moves were fathered by the agricultural colleges. Have the colleges changed their mind? Or is an Osage windbreak governed by a different kind of economics than a red pine windbreak?

There is still another kind of community planting where the thing to be planted is not trees but thoughts. To describe it, I want to plant some thoughts about a bush. It is called a bog-birch.

I select it because it is such a mousy, unobtrusive, inconspicuous, uninteresting little bush. You may have it in your marsh but have never noticed it. It bears no flower that you would recognize as such, no fruit which bird or beast could eat. It doesn't grow into a tree which you could use. It does no harm, no good, it doesn't even turn color in fall. Altogether it is the perfect nonentity in bushes; the complete biological bore.

But is it? Once I was following the tracks of some starving deer. The tracks led from one bog-birch to another; the browsed tips showed that the deer were living on it, to the exclusion of scores of other kinds of bushes. Once in a blizzard I saw a flock of sharp-tailed grouse, unable to find their usual grain or weed seeds, eating bog-birch buds. They were fat.

Last summer the botanists of the University Arboretum came

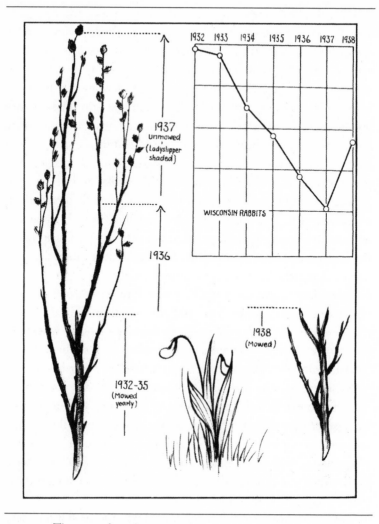

FIGURE 1: The story of a cycle

A mousy, unobtrusive, inconspicuous little bush, the bog-birch, plays an important role in the ups and downs of plant and animal life. Here is illustrated how it spells life or death to deer, grouse, rabbits, and ladyslippers in Wisconsin. In 1932 to 1935 rabbits were abundant and ate down the bog-birches each winter, giving the ladyslippers the sun. During 1936 and 1937 the cycle decimated the rabbits and the bog-birches grew high and shaded out the ladyslippers. In 1938 the rabbits recovered, mowed down the birches, and the ladyslippers regained their place in the sun.

to me in alarm. The brush, they said, was shading out the white ladyslippers in the Arboretum marsh. Would I ask the CCC crews to clear it? When I examined the ground, I found the offending brush was bog-birch. I cut the sample shown on the left of the drawing. Notice that up to two years ago rabbits had mowed it down each year. In 1936 and 1937 the rabbits had spared it, hence it grew up and shaded the ladyslippers. Why? Because of the cycle; there were no rabbits in 1936 and 1937. This past winter of 1938 the rabbits mowed off the bog-birch, as shown on the right of the drawing.

It appears, then, that our little nonentity, the bog-birch, is important after all. It spells life or death to deer, grouse, rabbits, ladyslippers. If, as some think, cycles are caused by sunspots, the bog-birch might even be regarded a sort of envoy for the solar system, dealing out appeasement to the rabbit, in the course of which a suppressed orchid finds its place in the sun.

The bog-birch is one of hundreds of creatures which the farmer looks at, or steps on, every day. There are 350 birds, ninety mammals, 150 fishes, seventy reptiles and amphibians, and a vastly greater number of plants and insects native to Wisconsin. Each state has a similar diversity of wild things.

Disregarding all those species too small or too obscure to be visible to the layman, there are still perhaps 500 whose lives we might know, but don't. I have translated one little scene out of the life-drama of one species. Each of the 500 has its own drama. The stage is the farm. The farmer walks among the players in all his daily tasks, but he seldom sees any drama, because he does not understand their language. Neither do I, save for a few lines here and there. Would it add anything to farm life if the farmer learned more of that language?

One of the self-imposed yokes we are casting off is the false idea that farm life is dull. What is the meaning of John Steuart Curry, Grant Wood, Thomas Benton? They are showing us drama in the red barn, the stark silo, the team heaving over the hill, the country store, black against the sunset. All I am saying is that there is also drama in every bush, if you can see it. When enough men know this, we need fear no indifference to the welfare of bushes, or birds, or soil, or trees. We shall then have no need of the word "conservation," for we shall have the thing itself.

The landscape of any farm is the owner's portrait of himself.

Conservation implies self-expression in that landscape, rather than blind compliance with economic dogma. What kinds of self-expression will one day be possible in the landscape of a cornbelt farm? What will conservation look like when transplanted from the convention hall to the fields and woods?

Begin with the creek: It will be unstraightened. The future farmer would no more mutilate his creek than his own face. If he has inherited a straightened creek, it will be "explained" to visitors, like a pock-mark or a wooden leg.

The creek banks are wooded and ungrazed. In the woods, young straight timber-bearing trees predominate, but there is also a sprinkling of hollow-limbed veterans left for the owls and squirrels, and of down logs left for the coons and fur-bearers. On the edge of the woods are a few wide-spreading hickories and walnuts for nutting. Many things are expected of this creek and its woods: cordwood, posts, and sawlogs; flood control, fishing, and swimming; nuts and wildflowers; fur and feather. Should it fail to yield an owl-hoot or a mess of quail on demand, or a bunch of sweet william or a coon-hunt in season, the matter will be cause for injured pride and family scrutiny, like a check marked "no funds."

Visitors when taken to the woods often ask, "Don't the owls eat your chickens?" Our farmer knows this is coming. For answer, he walks over to a leafy white oak and picks up one of the pellets dropped by the roosting owls. He shows the visitor how to tear apart the matted felt of mouse and rabbit fur, how to find inside the whitened skulls and teeth of the bird's prey. "See any chickens?" he asks. Then he explains that his owls are valuable to him, not only for killing mice, but for excluding other owls which *might* eat chickens. His owls get a few quail and many rabbits, but these, he thinks, can be spared.

The fields and pastures of this farm, like its sons and daughters, are a mixture of wild and tame attributes, all built on a foundation of good health. The health of the fields is their fertility. On the parlor wall, where the embroidered "God Bless Our Home" used to hang in exploitation days, hangs a chart of the farm's soil analyses. The farmer is proud that all his soil graphs point upward, that he has no check dams or terraces, and needs none. He speaks sympathetically of his neighbor who has the misfortune of harboring a gully, and who was forced to call in the CCC. The neighbor's check dams are a regrettable badge of awkward conduct, like a crutch.

Separating the fields are fencerows which represent a happy balance between gain in wildlife and loss in plowland. The fencerows are not cleaned yearly, neither are they allowed to grow indefinitely. In addition to bird song and scenery, quail and pheasants, they yield prairie flowers, wild grapes, raspberries, plums, hazelnuts, and here and there a hickory beyond the reach of the woodlot squirrels. It is a point of pride to use electric fences only for temporary enclosures.

Around the farmstead are historic oaks which are cherished with both pride and skill. That the June beetles did get one is remembered as a slip in pasture management not to be repeated. The farmer has opinions about the age of his oaks, and their relation to local history. It is a matter of neighborhood debate whose oaks are most clearly relics of oak-opening days, whether the healed scar on the base of one tree is the result of a prairie fire or a pioneer's trash pile.

Martin house and feeding station, wildflower bed and old orchard go with the farmstead as a matter of course. The old orchard yields some apples but mostly birds. The bird list for the farm is 161 species. One neighbor claims 165, but there is reason to suspect he is fudging. He drained his pond; how could he possibly have 165?

His pond is our farmer's special badge of distinction. Stock is allowed to water at one end only; the rest of the shore is fenced off for the ducks, rails, redwings, gallinules, and muskrats. Last spring, by judicious baiting and decoys, two hundred ducks were induced to rest there a full month. In August, yellow-legs use the bare mud of the water-gap. In September the pond yields an armful of water lilies. In the winter there is skating for the youngsters, and a neat dozen of rat-pelts for the boys' pin-money. The farmer remembers a contractor who once tried to talk drainage. Pondless farms, he says, were the fashion in those days; even the Agricultural College fell for the idea of making land by wasting water. But in the drouths of the thirties, when the wells went dry, everybody learned that water, like roads and schools, is community property. You can't hurry water down the creek without hurting the creek, the neighbors, and yourself.

The roadside fronting the farm is regarded as a refuge for the prairie flora: the educational museum where the soils and plants of pre-settlement days are preserved. When the professors from the college want a sample of virgin prairie soil, they know they can get it here. To keep this roadside in prairie, it is cleaned annually, always by burning, never by mowing or cutting. The farmer tells a funny story of a highway engineer who once started to grade the cutbanks all the way back to the fence. It developed that the poor engineer, despite his college education, had never learned the difference between a Silphium and a sunflower. He knew his sines and cosines, but he had never heard of the plant succession. He couldn't understand that to tear out all the prairie sod would convert the whole roadside into an eyesore of quack and thistle.

In the clover field fronting the road is a huge glacial erratic of pink granite. Every year, when the geology teacher brings her class out to look at it, our farmer tells how once, on a vacation trip, he matched a chip of the boulder to its parent ledge, two hundred miles to the north. This starts him on a little oration on glaciers; how the ice gave him not only the rock, but also the pond, and the gravel pit where the kingfisher and the bank swallows nest. He tells how a powder salesman once asked for permission to blow up the old rock "as a demonstration in modern methods." He does not have to explain his little joke to the children.

He is a reminiscent fellow, this farmer. Get him wound up and you will hear many a curious tidbit of rural history. He will tell you of the mad decade when they taught economics in the local kindergarten, but the college president couldn't tell a bluebird from a blue cohosh. Everybody worried about getting his share; nobody worried about doing his bit. One farm washed down the river, to be dredged out of the Mississippi at another farmer's expense. Tame crops were overproduced, but nobody had room for wild crops. "It's a wonder this farm came out of it without a concrete creek and a Chinese elm on the lawn." This is his whimsical way of describing the early fumblings for "conservation."

History of the Riley Game Cooperative, 1931–1939

In this piece, published in the Journal of Wildlife Management *in 1940, Leopold looks back over the nine-year history of the Riley Game Cooperative, described in an essay in part I, "Helping Ourselves." The cooperative, cofounded by Leopold, was a voluntary arrangement for city-dwelling hunters and rural farmers to manage a block of farmland for game, with the farmers providing the land, the hunters the cash, and everyone the needed labor. By and large, the cooperative was successful, although only because it became a university research project and graduate students stepped in to help run it. Much was learned at Riley about developing and restoring intensively used farmland to aid wildlife, and in this essay we see how Leopold put to use some of the European prac-*

tices that drew his praise in "Farm Game Management in Silesia." The Riley experience also highlighted a problem with the state's hunting preserve law: once the habitat was well developed and game populations became self-sustaining, there was no need to raise more captive birds, yet only by introducing birds could preserve owners get tags to hunt.

In 1940 THE Riley Game Cooperative will celebrate its ninth birthday.

Any farm game organization which promises to survive a decade is worth describing, for most of them die in infancy. Of some 350 started in the north central region since 1931, the survivors in 1936* could be counted on five fingers. The most vigorous survivals today are in Ohio.†

Of the six projects started in Wisconsin, two remain: Faville Grove‡ and Riley. This is the history of Riley.

Organization

One Sunday in the summer of 1931 I was cruising western Dane County, looking for a place to hunt during the approaching season. I stopped at a farmyard for a drink of water. The farmer, R. J. Paulson, was washing milk cans at the well. We talked game. He needed relief from trespassers who each year poached his birds despite his signs; I needed a place to try management as a means of building up something to hunt. We concluded that a group of farmers, working with a group of town sportsmen, offered the best defense against trespass, and also the best chance for building up game. Thus was Riley born.

*Aldo Leopold, "Farmer-Sportsmen Set-ups in the North Central Region," *Proc. North Amer. Wildlife Conf.* (1936), pp. 279–285.

†Lawrence E. Hicks, "The Controlled Hunting Areas and the Pheasant Refuge Management System of Northwestern Ohio," *Trans. 2d North Amer. Wildlife Conf.* (1937), pp. 589–598.

‡Arthur S. Hawkins, "A Wildlife History of Faville Grove, Wisconsin," *Trans. Wis. Acad. Sci., Arts & Letters* 32 (1940), pp. 29–65.

The Riley group now includes eleven farmers and five town members. The farm members furnish the land and also fencing, grain, and work. They have no cash outlay; the use of the land constitutes their "dues."

The town members furnish the operating funds and also help with the work.

All members share equally in the shooting, rules for which are drawn at an annual meeting. Current business is transacted by a committee of two "spokesmen," one for the farmers and one for the town group. There is no constitution, no by-laws.

The history of Riley has been affected by several events not foreseen at its inception; one was the passage of the Shooting Preserve Law soon after we organized in 1931. This law authorizes the Conservation Commission to license pheasant shooting preserves on which special seasons and bag limits may obtain. Instead of the usual ten-day season, a preserve may shoot from October to January. Instead of the usual bag limit of two cocks per day, a preserve may shoot, during the year, 75 per cent as many pheasants as it releases, without restriction as to sex. This season limit is enforced as follows: The local warden counts the birds released, and upon his certification the Department issues non-reusable tags, the number of tags being 75 per cent of the count. A tag must be attached to each bird killed. When breeders are released in spring, each hen counts as four prospective pheasants, and each cock as one.

Riley has been licensed as a preserve since 1931. The town members buy eggs and pay one of the farmers' wives for rearing the pheasants. We pay 50 cents per bird at eight weeks, at which time the warden makes the count and the birds are weighed and banded. After eight weeks the birds cease returning to the brooder coops and "go wild." Those killed during the shooting season are distinguished from birds of wild origin by their bands.

Another unforeseen event that affected Riley was the appointment of the author to the Chair of Wildlife Management, University of Wisconsin, in 1933. Since that time the area has served the dual purpose of a shooting ground for members and an experimental area for the university. Each year since 1936 a graduate student has been assigned to Riley. During weekends and vacations he makes censuses, conducts experiments in feeding and banding, and supervises plantings. He keeps all technical records. A complimentary shooting membership is voted to the student by the members, but he receives no other compensation.

Description

The Riley area is dairy country, closely grazed and cultivated, hence deficient in both winter food and winter cover. The average farm comprises 160 acres and maintains twenty-five cows and thirty-five hogs. Pastures occupy the creek bottoms which are too wet, and the ridges which are too stony, to plow. The only ungrazed cover consists of woodlots isolated by fields, bogs too soft for cattle to enter, and odd corners. To these has now been added a system of small fenced evergreen plantations or remises. The winter food, other than feeding stations, consists largely of crop residues (corn and soy-beans) accidentally left in the fields, manure spread on the fields, and ragweed aftermath on oat stubbles. The composition of the area by cover types is shown in Fig. 1 and in Appendix A.

The original vegetation was prairie, interspersed by groves of bur oaks. After farming had checked the free sweep of prairie fires, new stands of black and red oak seeded in on the prairie slopes and ridges. These post-settlement encroachments constitute the bulk of the present timber, which consists of scattered veterans of bur oak, branchy and open-grown, surrounded by even-aged black and red oaks grown in denser stands and dating from the Civil War or just before. Some roadsides still show relics of the prairie flora: prairie dock, lead plant, blazing star, butterfly weed, bluestem grass, and rattlesnake-master.

Riley lies at the foot of the terminal moraine of the Wisconsin glacier. Sugar Creek, the principal stream, drained the melting front of the ice sheet. The area is underlain by St. Peter's sandstone, which shows frequent outcrops, and by Galena dolomite. All the uplands are eroding, hence Sugar Creek is subject to sudden floods at all seasons.

Riley presents two major ecological problems: the gradual transfer of fertility from upland to bottoms by erosion, and the gradual elimination of cover by grazing.

This paper presents no remedy for the erosion problem. If the uplands grow too poor to raise corn, the farmers will be forced to ditch and cultivate the bottoms.

The cover deficiency is clearly increasing with time. Cutting in woodlots where cows prevent reproduction means the gradual elimination of woods. With each succeeding drouth, new marshes are plowed; their tussocks are cover of a sort, even when grazed bare. As uplands erode and are turned to pasture, the survival of fencerows becomes more difficult. Finally, since the CWA year

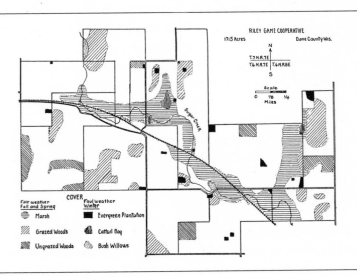

FIGURE 1: Cover map of Riley Game Cooperative

(1934) Riley has suffered a continuous loss of roadside cover through debrushing by highway crews. Before CWA debrushing was done every few years when needed, and helped wildlife by keeping the roadside in desirable shrubs and forbs. Today debrushing is done yearly or even twice yearly, and with constant regrading makes the roadside a refuge for quack grass.

The railroad right-of-way shows an opposite trend, and partly offsets the loss of cover on roadsides. It was formerly burned each year, but at our request is now cleared only when woody growths get too large. The railway strip is now the best cover on the area, and is closed as a refuge.

Food and Cover Improvements

The Riley Cooperative in its early years was a simple effort to increase game by quick and easy measures like restocking and winter feeding.

When Riley became a university project it expanded its aims and its audience and now deals with all wildlife, and with slow and difficult measures like planting cover. It aims to prove that the downward trend of wildlife in the dairy belt can be reversed by the combined efforts of farmers and sportsmen, without large expenditures either of cash or land.

Winter Feeding. Ten stations are operated from November to March. They require each year forty bushels of corn worth $25. Corn was at first bought and fed by the town members; it is now furnished and fed by the farmers. During the present winter a "boys' club" consisting of the sons of farm members has done most of the feeding.

Food patches supplemented by hoppers were tried at first. The patches were abandoned because it is the local custom to turn hogs loose on stubbles in fall. Hoppers are inconvenient because some of the farmers lack corn shellers. The feeding method now in favor is throwing ear corn on straw piles under shelters. Some of the ears are shelled into the straw to induce the birds to exercise. Some farmers feed the weed seeds that gather under corn shredders. Such seeds are excellent, but must be dried to prevent heating.

We at first fed in both upland and lowland coverts. Experience showed that pheasants and quail gravitate to marshes with the onset of severe weather, hence we no longer maintain many upland stations. Later on, when the new plantations develop dense cover on uplands, it may be possible to hold birds there in winter.

Cover Plantings. Riley offers a considerable acreage of fair-weather cover good for spring and fall use, but very little foul-weather cover efficient against deep snows and hard blizzards. The aim of the cover plantings is to bolster existing fair-weather coverts with spots of denser vegetation, and to create new dense coverts on areas now bare.

Fair-weather cover now occurs on about 300 acres, or 17 per cent of the area. It consists of grazed tussock marsh *(Carex stricta),* brushy ungrazed woodlots isolated by fields and too small to fence as pastures, and brushy grazed woodlots where for some unknown reason the hazel and gray dogwood brush has not been browsed out by cattle or choked out by sod.

Foul-weather cover now occurs on only 1 per cent of the area. It consists of cattail bogs too soft for cattle to enter, bush willow along streams and railroad tracks, and grape tangles or plum thickets in fencerows.

It was evident from the outset that doubling or trebling the 1 per cent of dense winter cover by plantings would give a much better balance between the two classes at a negligible cost in agricultural acreage. It was not evident, however, what to plant, where to plant, or how to plant. This we had to learn through experience.

Willow plantings for cover were tried in pastures as early as 1931, but all succumbed to browsing.

In 1934 a CWA crew fenced twenty-one units in which 1,300

plants were set out, mostly white and Norway spruce (2-0 stock), red cedar, grape, viburnums, and mulberry. All succumbed to the 1934 drouth. The weak fences built of scrap wire were all breached by cattle.

In 1936 a volunteer crew of students and farmers built five units of good fence. Fortunately no plantings were made, for this year brought the most intense drouth in local history.

In 1937 the Soil Conservation Service gave us a CCC crew to plant the new units, and to build and plant eight others totalling seven acres. Strong 2-2 and 2-3 stock was used, mostly red pine, and despite a considerable drouth the survival ran from 70 to 90 per cent.

In 1938 and 1939 the farmers did their own planting, built their own fences, and some even hoed the trees throughout the summer. Good rains fell in 1938. Despite a considerable drouth in 1939, losses were negligible. The farmers no longer want relief labor crews, for the annual "tree planting bee" has become an enjoyable social event.

These cover plantings constitute a remise system* which will double the area of foul-weather or true winter cover. The cover pattern, including remises, appears in Fig. 1. The planted remises are not yet large enough to function as cover, but the oldest will come into service during the next few years. The test of their success will be whether a further increase in kill occurs during this period (see Fig. 2).

The decade of experience may be summed up as follows:

1. Riley now has eighteen fenced remises of young evergreens, mostly red or Norway pine, aggregating twelve acres. Most of these plots include extra area for grass cover, and for later plantings. Fenced remises will eventually be increased to cover 2 per cent of the area.

2. The most promising remises are those in which strong 2-2 or 2-3 red pines were planted on sandstone hills. The farmers are will-

*Aldo Leopold, "Farm Game Management in Silesia," *Amer. Wildlife* 25, No. 5 (September–October 1936), pp. 67–68, 74–76.

ing to give these sites because they are too dry for pasture and too rocky for plowland. Wild stands of red pine occur on similar land at Pine Bluff, about five miles from Riley. The trees are planted 6′ x 6′ on 3′ scalps, and are hoed for one or two years. They make cover the fifth year.

3. On limestone (dolomite) outcrops where red pine does not thrive, red cedar is used. There is no cheap source of red cedar, for forest nurseries do not grow it. They should, for cedar post plantations would have economic as well as wildlife value.

4. Conifer plantings on lowlands, despite superior soil, are of doubtful practicability because of rabbit damage and choking by sod. Of the conifers, white spruce makes the best growth, but it has no chance against rabbits unless very large stock is used, or unless each tree is screened by a woven wire cylinder. Willows seem to be the most practicable planted cover for lowlands. Royal or yellow willow starts best from cuttings, and can be kept bushy by pollarding.

5. Rabbits are a threat to all conifers, even red pine, if placed in cover. Plantings in the open are seldom attacked. Red cedar is not damaged.

6. All plantings must be fenced, and only strong, durable fences will do. We are now planting red cedars between each pair of posts to assure a good fence in future years.

Pheasant Management

Stocking. Save for one stray cock seen in 1930, Riley was devoid of pheasants until we stocked the area in 1931. Shooting began in 1932. Stocking operations are summarized in Table 1. Since 1934 the annual August release has been around 100 birds eight weeks old. Since 1937 an additional small spring release has been made.

Pheasant Kill and Drift. The kill shows a steady increase through the decade. Its composition appears in Table 1 and its trend in Fig. 2.

TABLE I: Pheasant stocking and kill

Year	Releases Adults (Spring)	Young (August)	Legal Quota of Tags Number	Member	Kill Cocks	Hens	Total	Per cent tags used	Number of banded releases in kill
1931		24	18		0	0	0		x
1932		70	56	6?	14	4	18	32	x
1933		0	0	0	4	1	5[1]	x	x
1934		119	91	6	31	21	52	57	x
1935		110	82	5–7[2]	28	18	46	56	x
1936		70	52	3–4	13	7	20	38	1?
1937	12	125	122	6–9	40	12	52	43	4?
1938	15	130	130	8	49	16	65	50	12
1939	12	99	84	5–12	48	29	77	92	7
Total	39	747	635	x	227	108	335	x	24
Average	x	83	71	x	25	12	37	53	x

[1] Tags carried over from previous year; this practice is no longer permitted.

[2] Quotas lumped on December 1; some members killed up to the second figure.

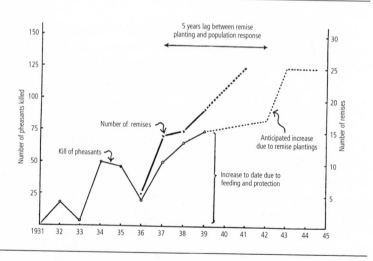

FIGURE I: Kill of pheasants and number of remises, Riley Game Cooperative

The hunting pressure has remained quite uniform. The same men and often the same dogs hunt for the same period year after year. Some are poor hunters and get little or nothing after the first week or two. Some are good hunters, shoot cocks only, and can bag the wildest cock if there is one left on the area. The net result has been a decade of hunting of uniform and normal intensity.

The pheasants respond to the combination of shooting, cold weather, and thinning cover by leaving the area. After the first week or two the season is closed outside, and this doubtless augments centrifugal drift. The tendency to drift outward, however, has decreased year by year; at first we were unable to kill more than a third of our allotment of birds; in 1939 we could, for the first time, have killed our full legal quota. The legal quota has been around 100 birds since 1934. One reason for the decreasing drift is probably the increasing population pressure outside. Riley drift has stocked the township, large outside plantings having been made during only the last two years.

Some birds still leave in winter, hence no census of the winter population has been attempted. Where these birds drift to we do not know. Wherever they go, they seem to return in spring, for there are always more cocks crowing in spring than wintered at our stations.

Effectiveness of Release, Banding. Since there were no pheasants prior to 1931, and since the kill has increased steadily ever since, it is safe to conclude that the early releases, made on empty or nearly empty range, survived and were effective. The gradual establishment of pheasants in the surrounding region occurred before large releases were made there.

In 1936 we began to mark our releases, partly to detect their recapture in the kill, partly to trace the direction of the winter exodus. The bag of marked birds appears in Table 1, and is remarkable for its almost negligible proportions.

In 1936 and 1937 the releases were marked by punching a hole in the web between the toes. Tests on confined birds showed that such

marks, unless made cleanly with a sharp punch, may close. Even if
the hole remains open, it may tear out and become unrecognizable,
or it may be overlooked. Hence the low returns from toe-punching
(1 to 4 birds per year) are clouded by doubt as to their complete-
ness.

In 1938 and 1939 releases were marked by colored celluloid and
also aluminum leg bands, or, if the bird was too small to carry
bands, by aluminum clips inserted in the web of the wing. Table 1
shows that of 214 releases thus marked during the last two years,
only 19 (9 per cent) were bagged on the area. We were able to trace
another dozen taken outside. Of 142 birds killed on the area since
reliable marking began, only 13 per cent were artificially propagated.

We must conclude, then, that releases no longer count for much
in the Riley bag; our steadily mounting kill comes not from current
releases, but from the increasing wild population.

The fact that as many bands are returned from outside as from
inside the area suggests that pen-raised pheasants may be more sus-
ceptible to dispersion than are wild birds. This suggestion is supported by the fact that large groups of 20 or more bearing colored
leg bands are seen on roads inside the area as late as September,
whereas the October bags show as many outside as inside. It is also
worthy of note that the instances of extreme dispersion (twenty to
forty miles) reported yearly in the sporting literature usually represent game-farm ringnecks, not wild pheasants.

Pheasant Economics. During the last two years the kill of 142
pheasants has cost us $200 in cash, or $1.40 per bird. This is well
under the current price for shooting on commercial preserves,
which runs from $2.50 to $3.00. However, our town members paid
all the cash costs, but killed only two-thirds of the total bag, hence
the cost of their take of birds was about $2.00 per pheasant.

That the licensed shooting preserve is an effective device for
establishing pheasants in new regions is indicated by the history
already given. It is not, however, an economical device for sus-

tained yield from established populations. The restocking operations required by the shooting preserve law account for 70 per cent of our yearly costs at Riley, but bring us only 13 per cent of our bag. Biologically speaking, the state could absolve us of this cost, provided that it required us to prove, by suitable records, that we had created a sustained wild yield, and provided it required us, by tags, to stay within the proved capacity for such yield. That capacity at this time is the total kill (about 75) minus the number of released birds (about 10) or about 65 pheasants. Such recognition of wild production would greatly cheapen unit costs, and thus encourage the establishment of more preserves and more food and cover restoration.

Would such a policy lighten the hunting pressure on outside range? Riley's sixteen memberships represent about twenty-five shooters, who use 1,700 acres; i.e., 70 acres each. The outside hunter doubtless has less room. The probable effect of encouraging wild-production preserves would be to increase the total pheasant population and the total bag, to decrease the hunting pressure in time (because of long seasons), but to increase the hunting pressure in space by reason of outside crowding.

Other Species

Bobwhite. The Riley quail population mounted steadily from about 40 in 1931 to about 180 in 1935. Quail were then nearly wiped out by the killing winter of 1935–1936. We have good censuses for subsequent Novembers: 1936, 65 quail; 1937, 53 quail; 1938, no record; 1939, 66 quail.

The Riley farmers refuse to permit quail shooting, even in years like 1935 when the season was open and the supply abundant.

No visible conflict between quail and pheasants has so far appeared.

Hungarian Partridge. This species had never been seen at Riley until 1937. In October of that year a wandering covey, drifting from the east by natural spread,* appeared. In 1938 mated pairs were seen, and a hunter killed 4 from a covey just outside the boundary. In 1939, pairs, nests, and several coveys of young were seen. The advent of partridges promises new production from the large areas of upland too bare to carry quail or pheasants.

Ruffed Grouse. A remnant which seems never to exceed 3 birds has inhabited the ungrazed woods in the southwest corner since 1931.

Prairie Chicken. A flock appeared each winter from 1932 to 1935, but none have been seen since. There is no recent record of breeding.

Cottontail. These are abundant and remarkably stable in numbers. The flushing rates noted in my journal since 1936 show 48 rabbits in forty-two hours, or 1.2 per hour during the hunting season. This index represents a population great enough to wreck plantations, and to consume a large fraction of the feed at stations.

There is a good deal of unauthorized boundary trespass by outside rabbit-hunters. This is overlooked as long as no birds are molested.

Waterfowl. Ducks are occasionally seen on Sugar Creek during migration. In wet years the creek bottom pastures are excellent jacksnipe ground. A great gathering of snipes occurred in the fall of 1936. An intense drouth first caused overgrazing of all the pastures, and heavy fall rains then soaked and flooded the trampled soil. On October 16, 1936, I flushed 50 snipes in thirty minutes. Probably 150 were shot that year. In the average year I estimate the kill of snipes at 50 birds.

In drouth years when Sugar Creek recedes and leaves bare muddy banks, it is visited by woodcocks. This occurred in August of 1936 and September, 1939. None are known to breed.

Predators. Our members account for perhaps a half-dozen feral

---

*Aldo Leopold, "Spread of the Hungarian Partridge in Wisconsin," *Trans. Wis. Acad. Sci., Arts & Letters* 32 (1940), pp. 5–28.

cats per year. Only one raptor has ever been killed by a member to my knowledge: a Cooper's hawk shot from a quail feeder in 1936.

Gray foxes are plentiful on the wooded hill in the southwest corner. In 1938, four grays were taken there by one pack of dogs. Foxes now seem to be spreading over the whole area; this may be a response to the increased pheasant crop, or to the 1938–1939 high in mice.

Minks, muskrats, and a few raccoons occur along Sugar Creek. In 1938 two coons were taken by farmers; in 1939 none were taken.

In general, predator control at Riley has been casual, and in respect of hawks and owls, non-existent. Despite this, a steadily increasing head of game has been built up. Riley, then, is proof that in this region predator control is unnecessary.

Fish. Sugar Creek was formerly a good trout stream. It is supposed that increasing floods have now eliminated all but a few German browns.

One Riley farmer has a small spring-fed slough covering a quarter of an acre that is stocked with mud minnows. Bait dealers paid him $20 in 1938 and $21 in 1939 for seining privileges in this slough, a remarkably high economic return from so small a water.

People

No farmer-sportsman group is stronger than the ties of mutual confidence and enthusiasm which bind its members. There was a time, about midway of its career, when enthusiasm at Riley faltered. We had lost two successive cover plantings through drouth, cattle breached our remise fences, we had some bad hatchings of pheasant eggs, and we suffered trespass from outside hunters who invaded the area via the railroad track. The organization nearly collapsed.

Riley was revived by Ellwood B. Moore and H. Albert Hochbaum, graduate students assigned to the area in 1936 and 1937, respectively. They rekindled enthusiasm, organized "planting bees," induced the railroad to post its tracks, and started an exchange of social events between the farm and town groups. Today the organization is very much alive. The annual Riley dinner is anticipated by the members quite as keenly as opening day on pheasants.

When hunters are taken out of the "devil-take-the-hindmost" atmosphere of free public shooting and become dependent upon their own efforts for sport, their attitude undergoes a remarkable shift toward conservatism. Thus at Riley the rules permit hens to be shot, but few members do it; in 1938 they carried their "chivalry" to such lengths that we had to buy cocks to rebalance the sex ratio. The rules permit a member to shoot his entire allotment of pheasants in one day, but the members seldom shoot more than one bird.

Some of the farmers have developed interests that extend far beyond game. Two of them, on their own initiative, have started an artificially planted tamarack grove, with the ultimate objective of reintroducing ladyslippers.

Riley has a waiting list of adjoining farmers who "want in." We have declined these expansions, believing that too large a membership would destroy the informality which is probably the key to our survival.

Summary

Riley is a farmer-sportsman shooting preserve for which farm members furnish the land and town members the cash; both contribute labor and share in the shooting. The university uses the area for wildlife studies.

The seasonal kill on 1,700 acres is now 75 pheasants. The area is fed and restocked, but there is no predator control. Banding shows that wild birds comprise the bulk of the kill, but restocking absorbs the bulk of the costs. The cost per bird in bag is $1.40. The pheasants disperse in fall and return in spring. Dispersion is decreasing.

If the law allowed, Riley could now abandon artificial restocking and kill 65 pheasants per year at 50 cents per bird in bag.

Planted remises of red pine cover 1 per cent of the area and furnish cover the fifth year. There is another 1 per cent of natural winter cover.

Appendix A: Vegetative Types in 1939

	Acres	Per Cent
Plowland		
Corn	219	13
Alfalfa	177	10
Small grains	234	14
Soy-beans	24	1
Subtotal	654	38
Grassland		
Pasture	650	38
Haymeadow	114	7
Ungrazed	7	0
Subtotal	771	45
Woodland		
Pasture	222	13
Ungrazed woods	31	2
Ungrazed brush	25	1
Fenced plantings	12	1
Subtotal	290	17
Grand total	1,715	100

Appendix B: Rules for 1939

Adopted at the Annual Meeting, October 12.

Open Season. October 14 to January 1.

Cock Rule. Either cocks or hens, but only 1 hen may be shot for each 2 cocks shot. After December 1 the spokesmen may modify the cock rule if they think conditions warrant.

Guests. Hunting privileges are limited to members and their immediate families. On rented farms the renter and owner together constitute one member.

Areas Closed as Refuges. (1) Right-of-way of C. & N. W. R. R.; (2) spring-hole on Paulson farm north of R. R. crossing; (3) fence-row between McCaughey and Brannan; (4) within 100 yards of any feeder containing feed.

Entirely Protected Species. Hungarian partridge, quail, ruffed grouse, prairie chicken.

Allotment. We have a total credit of 111 birds this year, 99 for young pheasants raised, 12 for cocks released last spring. Seventy-five per cent of 111 gives us a possible kill of 84 birds.

We have 16 members, plus one complimentary membership for the student, total 17. This allows 5 birds per member. After December 1, unused quotas will be lumped, and any member may shoot up to a season total of 12 birds.

Registration and Tags. All tags will be kept at Paulson's. All pheasants killed should be promptly tagged, registered, and weighed. All releases are banded. If you kill a banded bird, please register the number.

Identification of Members. To identify each other while hunting, members will raise the right hand when sighting other members.

APPENDIX C: BUDGET FOR A TYPICAL YEAR

	Expenditures	Receipts
5 town members at $20 each		$100.00
200 eggs for propagation	$20.00	
Rearing 100 pheasants at 50¢ (8 wks. old)	$50.00	
Shooting-preserve license	$10.00	
10 metal signs for posting at 30¢	$3.00	
Annual membership dinner	$17.00	
Total	$100.00	$100.00

APPENDIX D: PERSONNEL

Farm Members: R. J. Paulson, Joe and Jerome Brown, O. M. Hub, M. G. Thompson, L. C. England, Wesley Riley, Hillery McCaughey, Albert Bohle, J. L. Henderson, William Cook, Joe L. Brannan.

Town Members: T. E. Coleman, A. W. Schorger, Howard F. Weiss, R. J. Roark, Aldo Leopold.

Students: 1936: Ellwood B. Moore; 1937: H. Albert Hochbaum; 1938: Lyle K. Sowls; 1939: Bruce P. Stollberg.

Planning for Wildlife

This essay, written in 1941 but never published, shows how far Leopold had shifted in his understanding of wildlife conservation. As he had come to see it, wildlife conservation was no longer an independent aim; it was an essential means of promoting the well-being of "the land-mechanism" and adding to "the satisfactions of living." In the long term, he observes here, human welfare depends on a healthy land, complete with all or nearly all of its biological parts. This essay is also remarkable in presenting wildlife conservation as a positive duty that society might rightly impose on landowners, fairly but firmly. The farmer must be willing to devote part of his farm acreage to wildlife, "to suffer losses" in order to avoid eliminating native species, and to retain the fertility of his soil. "Wildlife education," Leopold tells us, "is no separate thing; it is part and parcel of land-education, and of social philosophy."

M*otives.* No one can write a plan for accomplishing something until the reasons for desiring to accomplish it are defined. The reasons for restoring wildlife are two:

1. It adds to the satisfactions of living.
2. Wild plants and animals are parts of the land-mechanism, and cannot safely be dispensed with.

The land-mechanism, like any other mechanism, gets out of order. Abnormal erosion, loss of soil fertility, excessive floods and drouths, the spread of plant and animal pests, the replacement of useful by useless vegetation, and the dying out of protected species are all disorders of the land-mechanism.

Science understands these disorders superficially, but it seldom understands why they occur. Science, in short, has subjugated land, but it does not yet understand why some lands get out of order, others not. Stable (i.e., healthy) land is essential to human welfare. Therefore it is unwise to discard any part of the land-mechanism which can be kept in existence by care and forethought. These parts might later be found to contribute to the stability of land. Most lands were stable before they were subjugated.

Many other motives have been asserted: economic profit, services to agriculture, stimulation of tourist business, etc. These hold good for some kinds of wildlife in some spots, but they break down in others. There is no economic profit in a ladyslipper. A peregrine falcon is detrimental to agriculture in every direct sense, but nevertheless worth conserving. A pheasant attracts more tourists than a prairie chicken, but has a far lesser value; he is not part of the native land-mechanism.

Man and Wildlife. The impact of civilization destroys many species of wildlife, some unavoidably (buffalo), many without any real reason (woods wildflowers). It greatly increases others, but a high proportion of these become pests, either native (rodents) or imported (carp, starling). It evicts species from one habitat and encourages them in others (shift of deer from the prairie border to the north woods).

The net result of these changes is a wild fauna and flora constantly decreasing in variety of species, in stability of populations, and in the ratio of benefits to damages. Another net result is a constantly increasing dependence on artificial replenishment from hatcheries and propagating plants, and on artificial control of "undesirable" species. Artificial replenishment and control are always costly and often ineffective.

The plan-wise adjustment of the impact of civilization can mitigate the losses and enlarge the gains in wildlife, and reduce the need for artificial interference.

Essentials of a Plan. The plan-wise adjustment is not primarily a matter of laws, appropriations, or administrative devices, but rather of modifying land-use so as to provide the habitat needed by each species. Hence the execution of a plan rests with farmers and landowners, rather than with government. The function of government is to teach, lead, and encourage.

The average farm has, or could have, a hundred resident bird species, a score of mammals, and several hundred plants. Of this total, perhaps a quarter will persist or disappear according to whether they are encouraged or ignored. These threatened species are usually the most interesting, useful, or beautiful ones. Each species has its own habitat requirements. Hence the retention of a rich fauna and flora is a rather complex job of habitat-engineering.

It does not suffice for the farmer to be interested in only one group of species. Exclusive interest in shootable game, for example, often means the deliberate extermination of the equally valuable predators. Exclusive interest in non-shootable wildlife means the needless elimination of wholesome sport. Exclusive interest in forests has, in parts of Europe, eliminated most other wildlife and ultimately damaged the forests themselves.

To retain any large fraction of his potential wildlife, the farmer must be willing to use odds and ends of land for special kinds of food and cover, and for water-retention. Two or three per cent of the farm acreage thus devoted to wildlife, plus the waste corners present on most farms, and crop residues present on all farms, often spells the difference between wildlife riches and poverty.

The farmer must also be willing to suffer losses, within reason, rather than eliminate a species from his community. The most useful hawks and owls, for example, occasionally take poultry; the finest songbirds take fruit; game birds eat grain.

Above all, the farmer must retain the fertility of his soil, for a rich fauna and flora, like a bountiful crop, is the direct expression of a rich and vigorous soil.

It is apparent that to accomplish these complex adjustments (which we may call collectively "wildlife management") the farmer must be moved by something more than a vague liking for wild things. He must be moved by a positive affection for the fauna and flora as a whole, and he must take pride in the skill and knowledge exercised in their management. In short, each farmer must build up, and cherish, his social "rating" as a producer of wild as well as tame animals and plants.

It is apparent, likewise, that the harvesting of game, fish, and fur crops must be the prerogative of the landowner, rather than the prerogative of the self-invited public. "Free shooting" in the end means nothing to shoot.

Wild Lands. Wildlife restoration on forests and ranges, as distinguished from farms, is likewise a matter of adjustments in land-use. These lands, being cheap, will have a higher proportion of public ownership, and hence a greater chance for quick governmental action as distinguished from slow landowner education. To offset this, they will have a higher proportion of public use, and the public always abuses common property. Hence education must be aimed primarily at the public.

On wild land, wildlife management can (in fact must) employ the most inexpensive means to guide natural processes, i.e., must rely on biological skill rather than on dollars, work, or legislative edicts. For example, in seeking to control the excess deer which are now spoiling their own range in many localities we rely entirely on guns, i.e., on legislative policy. We prefer to control deer by shooting, yet experience shows that guns always underdo or overdo the job, i.e., the method is ineffective. It seems probable that the natural predators which once stabilized all deer herds must be reintroduced and managed to supplement the function of the guns.

Again, in seeking to control the excess rodents which now retard the recovery of overgrazed ranges, we rely entirely on poisoning; i.e., on dollars and work. The method is ineffective because it must be constantly repeated, costs are high, and many useful animals are killed. It seems probable that rodents cannot be controlled by poison or traps alone. They are probably themselves an expression of overgrazing, and can be controlled only by restoring the plants least useful to them, and most useful to livestock.

Completely wild lands have one function which is important, but as yet ill-understood. Every region should retain representative samples of its original or wilderness condition, to serve science as a sample of normality. Just as doctors must study healthy people to understand disease, so must the land sciences study the wilderness to understand disorders of the land-mechanism.

Education. A wildlife restoration plan is thus a plan for educating landowners, private and public, to want wildlife, and to understand how their wants may be fulfilled. This may sound like propaganda, but the shoe is on the other foot. We must *undo* the propaganda, brought to bear on landowners for the last century, which teaches that the land is a factory to be operated solely for profit. The land is a factory, but it is also a place to live, and wildlife helps make it a good place.

What took a century to do cannot be undone in a decade. Education must begin at the bottom and work upward. The land-philosophy of agricultural schools and extension agencies must be turned inside out. Wildlife education is no separate thing; it is part and parcel of land-education, and of social philosophy.

For example: One of the common denominators of all land problems is the plant succession, i.e., the sequence of plant coverings on a given soil. All farming, all forestry, all gardening, all landscaping, and all wildlife management is a manipulation of this sequence. Botany, geography, geology, zoology, and even history can be understood only if the plant succession is understood. Yet how many "educated" persons, or even teachers, know the plant succession of their own back yards? Botany, as now taught, is the number of hairs on a leaf; zoology the innards of a frog; history the dates of battles. Why do the Germans covet the Ukraine? Because its prairie soils favor an annual grass, wheat, as the first stage in its plant succession.

This face-about in land philosophy cannot, in a democracy, be imposed on landowners from without, either by authority or by pressure groups. It can develop only from within, by self-persuasion, and by disillusionment with previous concepts. Shortcuts like conservation text books, and conservation programs in youth organizations, help if they are sound and honest, but they are microscopic fractions of a deep and slow process. A wildlife plan is a constantly shifting array of small moves, infinitely repeated, to give wildlife due representation in shaping of the future minds and future landscapes of America.

Biotic Land-Use

This previously unpublished essay, written around 1942, is a milestone in Leopold's growing grasp of land health and its practical implications. In it, Leopold explores how and why ecology should guide land-use practices. He criticizes practices aimed only at particular goals such as crop production, game management, and control of floods and erosion. Good land use, Leopold tells us, requires a single, coordinated goal. Here, Leopold terms that goal "stability," which he uses synonymously with "the

health of the land as a whole." Science is never likely to "write a formula"
for land stability, he predicts. Lacking that, the best measures of stability
are soil fertility, retention of native fauna and flora, and efficient recy-
cling of nutrients. Leopold may have written this essay to introduce a
book or some other never-completed study of land-use practices.

MANAGEMENT is conserving particular plants or animals by
keeping the land favorable.

Biotic land-use is conserving land by keeping the plants and
animals favorable.

The biotic idea is thus an *extension* of the idea of management,
and it asserts the *converse* of the management theorem.

Both stem from ecology. The biotic idea merely translates ecology for purposes of guiding land-use.

The term "land" includes soils, water systems, and wild and
tame plants and animals.

Conservation is the attempt to understand the interactions of
these components of land, and to guide their collective behavior
under human dominance.

Land-use problems are of two orders.

We hear most about those problems which hinge on questions
of supply and demand. Timber famine, agricultural adjustment,
and attempts to husband the supply of water and game are familiar
examples. This order is, for present purposes, not very important,
because it deals with visible forces which are amenable to social
controls. Solutions are possible, and ways and means are known.

We hear least about another order which is very important
because ways and means are not known, or are ineffective. It consists of dislocations of land which present no visible cause. Thus
some species irrupt as weeds or pests, while others disappear, both
without visible reason.

It also includes dislocations of land for which a cause is visible, but for which the social controls so far used are inadequate. Thus we know, at least superficially, what causes soil erosion and floods, but the present program can hardly be called a cure.

There are intermediate land problems. Thus forestry and range management, if applied, can raise wood and grass; agronomy can raise crops; these problems are of the first order. But the restoration of full soil health and productivity is another matter, and falls in the second order.

Thus we see that the *basic* problem in land-use, the problem which directly underlies the second order and indirectly the first, is the stability of the land mechanism.

Stability is characteristic of new land. Undisturbed communities change their composition and their internal economy only in geological time. Within the time-scale of human affairs, they are stable.

Another characteristic of new land is diversity. The biotic community is diverse in composition, complex in organization, and tends to become more so.

When the technologies are applied to land they achieve, each within its own field, various degrees of success. There is a tacit assumption that the sum of their successes equals stable land. It is assumed that if good agronomy, erosion control, flood control, pasture management, forestry, and wildlife management be simultaneously applied to a given area, stability will follow. It is admitted that this assumption is conditional upon something which technicians in khaki call "coordination." Planners in tweeds call it "integration."

Many efforts have been made to define and implement coordination, but I recall no effort to examine the validity of its basic premise. Is it true that, given good coordination, the sum of the technologies equals stable land?

It is common knowledge that the technologies are partially competitive. "Good" agronomy means a coverless countryside devoid of all but the least exacting wild species. It means widespread drainage with possible derangement of water systems; it means the extinction of marsh and bog communities. "Good" pasture management relegates woods to the poorest slopes. "Good" forestry, until very recently, meant artificial monotypes which excluded wildlife and sometimes sickened the soil itself. "Good" game management, in Europe at least, abolishes the predators, which is in turn presumably accountable for irruptions of rodents. Can good coordination iron out these conflicts? Perhaps, but it seems safe to say that there are no instances in which it has yet done so.

The technologies are usually applied too late, and they are seldom all applied with equal intelligence to an entire land unit. We have no evidence on what they could do if perfectly balanced and timed. Our only guide is their collective performance, so far, on those land units where the largest number of them have been applied, for the longest time, with the most earnest attempt at coordination. In America, most such attempts are so far governmental rather than private. I will cite two as examples.

In southwestern Wisconsin erosion control, flood control, pasture renovation, crop rotation, nitrification by legumes, woodlot improvement, and wildlife management have been applied for a decade. Each has scored its own success in spots, and the disorganization of the land has doubtless been reduced in its velocity. But this region still displays flashy streams, loss of topsoil, silting of reservoirs, migration of plowland from upland to marshes and flood-channels, irruption of white grubs and weed pests, exaggerated drouth damage, falling water table, and scarcity of upland game. The momentum of erosion started during the wheat era and the dairy boom is certainly reduced, probably not arrested, certainly not reversed. It seems doubtful whether the sum of the technologies will stabilize this land.

Again: In the Southwest, erosion control, range management, stock water development, reclamation by irrigation and pumping, and mountain forestry have been applied during a period varying from ten to forty years. Each has scored its own success in spots, notably national forestry and range management on the headwaters. But they do not add up to stable land. All came too late, after erosion due to early overgrazing had gained momentum. The result: silted reservoirs, tearing out of valleys, widespread drainage of already dry soils by gullies, wholesale conversion of grass to chaparral, wholesale replacement of palatable by unpalatable range plants, irruption of rodent pests, loss of vulnerable and predacious wild species, falling water tables, dust storms. This land was set on a hair-trigger, and it seems doubtful whether the sum of the technologies will ever reclaim it. The disease will run its course and end up in new, and lower, levels of productivity.

These are only two instances. As evidence in the court of land science, both are defective in that technology came too late. But a glance at world experience indicates that technology *usually* comes too late. It seems academic, therefore, to say (as I myself have done) that the technologies are preventatives, not cures, and that applied in time, they will successfully preserve for land its normal stability of organization, or health. It seems more realistic to conclude that conservation, at bottom, is not to be accomplished by any mere mustering of technologies. Conservation calls for something which the technologies, individually and collectively, now lack.

What do they now lack? At this point I perforce depart from scientific logic, for we are beyond the range of scientific evidence. What I offer is opinion, or, if you prefer, judgment.

They lack, firstly, a collective purpose: stabilization of land as a whole. Until the technologies accept as their common purpose the health of the land as a whole, "coordination" is mere window-dressing, and each will continue in part to cancel the other. The acceptance of this common purpose does not call for the surrender of their separate purposes (soil, timber, game, etc.) except as these conflict with the common one.

They lack, secondly, a collective yardstick for appraising ways and means to stabilization or land-health. Each technology has its own yardsticks, usually yields or profits. But only commercial land-uses have any profits, and some of the most important land-uses have only spiritual and esthetic yields. The collective criterion of good land-use must be something deeper and more important than either profit or yield. What?

Among the ordinary yardsticks, I can think of but one which is obviously a common denominator of success in all technologies: soil fertility. That the maintenance of at least the original fertility is essential to land-health is now a truism, and needs no further discussion.

What else? What, in the evolutionary history of this flowering earth, is most closely associated with stability? The answer, to my mind, is clear: diversity of fauna and flora.

It seems improbable that science can ever analyze stability and write a formula for it. The best we can do is to recognize and cultivate the general conditions which seem to be conducive to it. Stability and diversity are associated. Both are the end-result of evolution to date. To what extent are they interdependent? Can we retain stability in used land without retaining diversity also?

There are two ways to explore this question: Examine the performance of lands where diversity has been lost, and examine the land mechanism itself for leads.

Northwestern Europe is the only part of the globe presenting an intensively used landscape which seems to have remained stable despite the loss of diversity in its fauna and flora. That its farm soils remain fertile is well known, and this alone is an achievement of world-wide importance. Part of its forest soils are sick, but the reason is now known and in process of correction. Its water systems, despite ruthless artificialization, still produce many fish, few floods, and little silt. Pest-like irruptions of plants and animals occur, but the pest problem, save for rabbits and forest insects, is perhaps less serious than with us. The non-game fauna has lost its large carnivores, but the game fauna, save for large mammals, is intact. The flora shows severe shrinkages and possibly some early and unrecorded extinctions. The migratory game birds are in a bad way, and are maintained only by replenishment from the Asiatic reservoir. The small bird fauna is warped in its composition, and some migratory species are threatened, but this is chargeable to the Latin nations where they winter.

By and large, this part of Europe has not lost its stability. Man himself is here far less stable than his land. That the present human dislocations are, at bottom, the expression of an ever-artificialized ecological mechanism is probable, but beyond the scope of this paper.

The question in hand is whether other parts of the globe can remain stable without the deliberate retention of diversity. All I can say is that I doubt it. Land is unequally sensitive. All other parts of the globe are either undeveloped (the tropics, the arctics), in process of dislocation (most of United States, South Africa, Australia, China), or already relapsed into a retrograded stability (Mediterranean countries).

What leads can we derive from the land mechanism itself?

No "language" adequate for portraying it exists in any science or art, save only ecology. A language is imperative, for if we are to guide land-use we must talk sense to farmer and economist, pioneer and poet, stockman and philosopher, lumberjack and geographer, engineer and historian.

The ecological concept is, I think, translatable into common speech.

A rock decays and forms soil. In the soil grows an oak, which bears an acorn, which feeds a squirrel, which feeds an Indian, who lays him down in his last sleep to grow another oak.

This sequence of stages in the transmission of food is a food chain. It is a fixed route or channel, established by evolution. Each link is adapted to extract food from the preceding link and hand it on to the succeeding one. The links form a circuit.

The chain is not a closed circuit. Squirrels do not get all the acorns, nor do Indians get all the squirrels; some return directly to the soil. The food channel leaks at every link; only part of the food reaches its terminus.

Food is likewise sidetracked into branch chains. Thus the squirrel drops a crumb of his acorn, which feeds a quail, which feeds a horned owl, which feeds a parasite. This chain branches like a tree.

The owl eats not only quail, but also rabbit, which is a link in another chain: soil—sumac—rabbit—tularemia. The rabbit eats a hundred other shrubs and herbs. Each animal and plant is the intersection of as many chains as there are species in its dietary. The whole system is cross-connected.

Nor is food the only link which connects them. The oak grows not only acorns, it grows fuel, browse, hollow dens, leaves, and shade on which many species depend for food and cover or other services. The chains are not only food chains, they are chains of dependency for a maze of services, competitions, piracies, and cooperations. This maze is complex; no living man can blueprint the biotic organization of a single acre, yet the organization is clearly there, else the member species would disappear. They do not disappear. Fossil bones and pollens tell us that our fauna and flora remained virtually intact since the ice age, which is 200 centuries.

Soil, the repository of food between its successive trips through the chains, tends to wash downhill, but this downhill movement is slow, and in healthy land is offset by the decomposition of rocks. Some animals likewise accomplish an uphill movement of food.

Stability is the continuity of this organized circulatory system. Land is stable when its food chains are so organized as to be able to circulate the same food an indefinite number of times.

Stability implies not only characteristic kinds, but also characteristic numbers of each species in the food chains. Thus the characteristic number of the aboriginal Indian was small; more Indians would have killed each other or their hunting ground, less would have been blotted out by some blizzard, drouth, or epidemic.

We have now modified both the species-composition of the food chains and the characteristic numbers of their constituent species. Chains now begin with corn and alfalfa instead of oaks and bluestem. The food, instead of flowing into elk, deer, and Indians, flows into cows, hogs, and poultry; farmers, flappers, and freshmen. The remaining wildlife eats tame as well as wild plants.

These substitutions are, perforce, accompanied by readjustments. To every tinkering with every link in every food chain, the whole land mechanism responds with a readjustment. We do not understand or see them, for they usually occur without perceptible dislocations. We are unconscious of them, unless and until the end-effects turn out to be bad.

Along with the deliberate and beneficial substitutions come many accidental ones (Japanese beetle, creeping Jenny, Canada thistle, chestnut blight, blister rust), most of which are bad, some ruinous. Some sober ecologists predict that a few generalized plants and animals will ultimately usurp the whole globe.

The modified land mechanism, thus converted for human use, is often unstable—i.e., it can no longer recirculate the same food an indefinite number of times. Erosion, floods, pests, loss of species, and other land-troubles without visible cause are the expressions of this instability. Would the deliberate retention of both fertility and diversity reduce instability? I think it would. But I admit in the same breath that I can't prove it, nor disprove it. If the trouble is in the plant and animal pipelines, I think it would help to keep them more nearly intact. This is only a probability based on evolution, but it is the only help in sight.

The retention of fertility is already an accepted criterion of good land-use, in theory. It needs only conversion into practice.

There remains the question: Is there room, within the existing technologies, and without the disruption of our land economy, to retain more diversity of fauna and flora? Is it possible, by deliberate social effort, to reduce the frequency and violence of changes in the land mechanism, and still use the land?

At this point I digress to refute the notion, unhappily cultivated by ecologists, that the land mechanism has a kind of Dresden china delicacy, and falls to pieces at a loud noise. The whole history of civilization shows land to be tough. Lands differ in their toughness, but even the most sensitive took several generations of violence to spoil. The pioneer has always striven for violent, not gentle, conversion to human use, and most of the technologies, especially agriculture and engineering, are still uninhibited in this respect. In fact, wildlife management, and to some extent forestry, are the only technologies conscious of "naturalism" in land-use.

Return now to our question: Could the frequency and violence of land changes be reduced by deliberate social effort? Can changes already made be tempered in violence? Can the technologies agree on stabilization as their collective purpose, and on fertility and diversity as their yardsticks of progress? Do we ourselves, as a group, believe what we cannot prove: that retaining the diversity of our fauna and flora is conducive to stable land? These are the questions now to be discussed.

What Is a Weed?

In this essay, written in 1943 but published only in 1991, Leopold places a humorous veneer over his considerable frustration about the misguided advice given out by agricultural colleges and extension services. Nominally, it reviews a book on weeds issued by the Iowa Geological Survey, but for Leopold the book is merely a stand-in for a way of valuing nature that he finds ecologically uninformed and ethically, aesthetically, and historically impoverished. The book, he relates sadly, is "only one sample of a powerful propaganda, conducted by many farming states, often with the aid of federal subsidy." Perhaps Leopold left the piece unpublished because it struck too hard at his agricultural colleagues. The essay contains a blank space of several lines that Leopold never filled in.

To LIVE in harmony with plants is, or should be, the ideal of good agriculture. To call every plant a weed which cannot be fed to livestock or people is, I fear, the actual practice of agricultural colleges. I am led to this baleful conclusion by a recent perusal of *The Weed Flora of Iowa,* one of the authoritative works on the identification and control of weed pests.*

"Weeds do an enormous damage to the crops of Iowa" is the opening sentence of the book. Granted. "The need of a volume dealing with weeds . . . has long been felt by the public schools." I hope this is true. But among the weeds with which the public schools feel need of dealing are the following:

> Black-eyed Susan *(Rudbeckia hirta)* "succumbs readily to cultivation."

A model weed!

> Partridge pea *(Cassia chamaecrista)* "grows on clay banks and sandy fields," where it may be "readily destroyed by cutting."

The inference is that even clay banks must be kept clean of useless blooms. Nothing is said of the outstanding value of this plant as a wildlife food, or of its nitrogen-fixing function.

> Flowering spurge *(Euphorbia corollata)* is "common in gravelly soils" and "difficult to exterminate. To eradicate this plant the ground should be given a shallow plowing and the root-stocks exposed to the sun."

---

*The Weed Flora of Iowa, Bulletin No. 4 (Iowa Geological Survey, 1926), 715 pp.

Nothing is said of the wisdom of plowing gravelly soils at all, or of the fact that this spurge belonged to the prairie flora, and is one of the few common relics of Iowa's prairie years. Presumably the public schools are not interested in this.

Prairie goldenrod *(Solidago rigida),* which "though often a very troublesome weed in pastures, is easily killed by cultivation."

The locality troubled by this uncommon and lovely goldenrod is indeed exceptional. The University of Wisconsin Arboretum, in order to provide its botany classes with a few specimens to look at, had to propagate this relic of the prairie flora in a nursery. On my own farm it was extinct, so I hand-planted two specimens, and take pride in the fact that they have reproduced half a dozen new clumps.

Horsemint *(Monarda mollis).* "This weed is easily exterminated by cultivation," and "should not be allowed to produce seeds."

During an Iowa July, human courage, likewise, might easily be exterminated but for the heartening color-masses and fragrance of this common (and as far as I know) harmless survivor of the prairie days.

Ironweed *(Veronia baldwinii)* is "frequently a troublesome weed, but it is usually not difficult to exterminate in cultivated fields."

It would be difficult to exterminate from my mind the August landscape in which I took my first hunting trip, trailing after my father. The dried-up cowtracks in the black muck of an Iowa bottomland looked to me like small chasms, and the purple-topped ironweeds like tall trees. Presumably there are still school children who might have the same impressions, despite indoctrination by agricultural authority.

Peppermint *(Mentha piperita).* "This plant is frequently found along brooks. The effectual means of killing it is to clear the ground of the root-stocks by digging."

One is moved to ask whether, in Iowa, nothing useless has the right to grow along brooks. Indeed why not abolish the brook, which wastes many acres of otherwise useful farm land.

> Water pepper *(Polygonum hydropiper)* is "not very troublesome . . . except in low places. Fields that are badly infested should be plowed and drained."

No one can deny that this is a weed, albeit a pretty one. But even after drainage, would not some annual, and perhaps a more troublesome one, follow every plowing? Has Iowa repealed the plant succession? It is also of interest to note that the Iowa wildlife research unit finds *Polygonum hydropiper* to be [Several lines of blank space appear here in the manuscript. *Ed.*]

> Wild rose *(Rosa pratincola).* "This weed often persists," as a relic of the original prairie flora, "in grain fields of northern Iowa. Thorough cultivation for a few seasons will, however, usually destroy the weed."

No comment.

> Blue vervain *(Verbena hastata)* and hoary vervain *(V. stricta).* The vervains, admittedly weedy, are "easily destroyed by cultivation" and are "frequent in pastures," but nothing is said about why they are frequent.

The obvious reasons are soil depletion and overgrazing. To tell this plain ecological fact to farmers and school children would seem proper in an authoritative volume on weed control.

> Chicory *(Cichorium intybus)* "is not often seen in good farming districts except as a wayside weed. Individual plants may be destroyed by close cutting and applying salt to the root in hot dry weather."

School children might also be reminded that during the hot dry weather this tough immigrant is the only member of the botanical melting-pot courageous enough to decorate with ethereal blue the worst mistakes of realtors and engineers.

If the spirit and attitude of *The Weed Flora of Iowa* were peculiar to one book or one state, I would hardly feel impelled to challenge it. This publication is, however, only one sample of a powerful propaganda, conducted by many farming states, often with the aid of federal subsidy, and including not only publications but also weed laws and specialized extension workers. That such a propaganda is necessary to protect agriculture is, I think, obvious to all who have ever contended with a serious plant pest. What I challenge is not the propaganda, but the false premises which seem to be common to this and all other efforts to combat plant or animal pests.

The first false premise is that every wild species occasionally harmful to agriculture is, by reason of that fact, to be blacklisted for general persecution. It is ironic that agricultural science is now finding that some of the "worst" weed species perform useful or even indispensable functions. Thus the hated ragweed and the seemingly worthless horseweed are found to prepare the soil, by some still mysterious alchemy, for high-quality high-yield tobacco crops. Preliminary fallowing with these weeds is now recommended to farmers.*

The second false premise is the emphasis on weed control, as against weed prevention. It is obvious that most weed problems arise from overgrazing, soil exhaustion, and needless disturbance of more advanced successional stages, and that prevention of these misuses is the core of the problem. Yet they are seldom mentioned in weed literature.

These same false premises characterize public predator control. Because too many cougars or wolves were incompatible with livestock, it was assumed that no wolves or cougars would be ideal for

*W. M. Lunn, D. E. Brown, J. E. McMurtrey, Jr., and W. W. Garner, "Tobacco Following Bare and Natural Weed Fallow and Pure Stands of Certain Weeds," *Journal of Agricultural Research* 59, No. 11 (1939), pp. 829–846.

livestock. But the scourge of deer and elk which followed their removal on many ranges has simply transferred the role of pest from carnivore to herbivore. Thus we forget that no species is inherently a pest, and any species may become one.

The same false premises characterize rodent control. Overgrazing is probably the basic cause of some or most outbreaks of range rodents, the rodents thriving on the weeds which replace the weakened grasses. This relationship is still conjectural, and it is significant that no rodent-control agency has, to my knowledge, started any research to verify or refute it. Still if it is true, we may poison rodents till doomsday without effecting a cure. The only cure is range-restoration.

The same false premises beset the hawk and owl question. Originally rated as all "bad," their early defenders sought to remedy the situation by reclassifying part of them as "good." Hawk-haters, and gunners with a trigger-itch, have had lots of fun throwing this fallacy back in our faces. We should have been better off to assert, in the first place, that good and bad are attributes of numbers, not of species; that hawks and owls are members of the native fauna, and as such are entitled to share the land with us; that no man has the moral right to kill them except when sustaining injury.

It seems to me that both agriculture and conservation are in the process of inner conflict. Each has an ecological school of land-use, and what I may call an "iron heel" school. If it be a fact that the former is the truer, then both have a common problem of constructing an ecological land-practice. Thus, and not otherwise, will one cease to contradict the other. Thus, and not otherwise, will either prosper in the long run.

The Outlook for Farm Wildlife

Leopold contributed this essay in 1945 to the North American Wildlife Conference. It is a somber appraisal. His particular concern, reflecting his by then firm understanding of the land as community, was the "accelerating disorganization of those unknown controls which stabilize the flora and fauna, and which, in conjunction with stable soil and a normal regimen of water, constitute land-health." Leopold aptly contrasts the farm-as-food-factory philosophy with his own conservation view, in which "the criterion of success is a harmonious balance between plants, animals, and people; between the domestic and the wild; between utility and beauty." He also sees looming on the horizon even greater problems stemming from industrialized agriculture, citing DDT as an example.

TWENTY YEARS have passed since Herbert Stoddard, in Georgia, started the first management of wildlife based on research.

During those two decades management has become a profession with expanding personnel, techniques, research service, and funds. The colored pins of management activity puncture the map of almost every state.

Behind this rosy picture of progress, however, lie three fundamental weaknesses:

1. Wildlife habitat in fertile regions is being destroyed faster than it is being rebuilt.
2. Many imported and also native species exhibit pest behavior. A general disorganization of the wildlife community seems to be taking place.
3. Private initiative in wildlife management has grown very slowly.

In this appraisal of the outlook, I deal principally with the first two items in their bearing on farm wildlife.

Gains and Losses in Habitat. Wildlife in any settled country is a resultant of gains and losses in habitat. Stability, or equilibrium between gains and losses, is practically non-existent. The weakness in the present situation may be roughly described as follows: On worn-out soils we are gaining cover but losing food, at least in the qualitative sense. On fertile soils we are losing cover, hence the food which exists is largely unavailable.

Where cover and food still occur together on fertile soils, they often represent negligence or delay, rather than design.

There is a confusing element in the situation, for habitat in the process of going out often yields well.

For example, on the fertile soils of southern Wisconsin, the strongholds of our remaining wildlife are the woodlot, the fence-row, the marsh, the creek, and the cornshock. The woodlot is in process of conversion to pasture. The fencerow is in process of abolition; the remaining marsh is in process of drainage; the creeks are getting so flashy that there is a tendency to channelize them. The

cornshock has long been en route to the silo, and the corn borer is speeding the move.

Using pheasant as an example, such a landscape often yields well while in process of passing out. The marsh, grazed or drained or both, serves well enough for cover up to a certain point, while the manure-spreader substitutes for cornshocks up to a certain point. The rapid shift in the status of plant successions may in itself stimulate productivity.

The situation is complicated further by a "transmigration" of land use. Originally uplands were plowed and lowlands pastured. Now the uplands have eroded so badly that corn yields are unsatisfactory, hence corn must move to the lowlands while pasture must move to the uplands. In order that corn may move to the lowlands, they must be either tiled, drained, or channelized. This, of course, tends to destroy the remaining marsh and natural stream.

The upshot is a good "interim" crop which has a poor future. I don't know how widely a similar situation prevails outside my own state, but I suspect that the basic pattern, with local variations, is widely prevalent.

Runaway Populations. Wildlife is never destroyed except as the soil itself is destroyed; it is simply converted from one form to another. You cannot prevent soil from growing plants, nor can you prevent plants from feeding animals. The only question is: What kind of plants? What kind of animals? How many?

Ever since the settlement of the country, there has been a tendency for certain plants and animals to get out-of-hand. These runaway populations include weeds, pests, and disease organisms. Usually these runaways have been foreigners (like the carp, Norway rat, Canada thistle, chestnut blight, and white pine blister rust), but native species (like the June beetle and various range rodents) are clearly also capable of pest behavior.

Up to the time of the chestnut blight, these runaways did not threaten wildlife directly on any serious scale, but they now do, and it is now clear that the pest problem is developing several new and dangerous angles:

1. World-wide transport is carrying new "stowaways" to new habitats on an ascending scale. (Example: *Anopheles gambiae* to Brazil, bubonic plague to western states.)
2. Modern chemistry is developing controls which may be as dangerous as the pests themselves. (Example: DDT.)
3. Additional native species, heretofore law-abiding citizens of the flora and fauna, are exhibiting pest behavior. (Example: excess deer and elk.)

These three new angles must be considered together to appreciate their full import. Mildly dangerous pests like ordinary mosquitoes evoked control measures which severely damaged wildlife; desperately dangerous pests will evoke corresponding control measures, and when these collide with wildlife interests, our squeak of pain will not even be heard.

Moreover, wildlife itself is threatened directly by pests. Sometimes they hit so fast and hard that the funeral is over before the origin of the malady is known. Thus in Wisconsin, we have a new disease known as burn blight, the cause of which is still unknown. It threatens to destroy young Norway pine and jack pine, especially plantations. Oak wilt, the cause of which was only recently discovered, is steadily reducing red and black oaks. Our white pine is already blighted except on artificially controlled areas. Bud-worm is in the spruce. Hickory can't grow because of a weevil which bites the terminal bud. Deer have wiped out most white cedar and hemlock reproduction. Sawfly has again raided the tamaracks. June beetles began years ago to whittle down the bur and white oaks, and continue to do so. Bag worm is moving up from the south and west and may get our red cedars. Dutch elm disease is headed west from Ohio. What kind of a woodlot or forest fauna can we support if every important tree species has to be sprayed in order to live?

Shrubs are not quite so hard hit, but the shrub flora has its troubles. On the University of Wisconsin Arboretum, an area dedicated to the rebuilding of the original native landscape, the Siberian honeysuckle is calmly usurping the understory of all woods, and threatens to engulf even the marshes.

In Wisconsin woodlots it is becoming very difficult to get oak reproduction even when we fence out the cows. The cottontails won't let a young oak get by. One can't interest the farmer in a woodlot which reproduces only weed trees.

Of the dozen pests mentioned here, four are imported, seven are runaway native species, and one is of unknown origin. Of the twelve, six have become pests in the last few years.

Farm crops and livestock exhibit a parallel list of pests, of which the worst now rampant in my region is the corn borer. The corn borer can be controlled by fall plowing, but what that will do to cornbelt wildlife is something I dislike to think about.

It all makes a pattern. Runaway populations are piling up in numbers and severity. In the effort to rescue one value, we trample another. Wild plants and animals suffer worst because we can't spend much cash on controls or preventatives. Everything we lose will be replaced by something else, almost invariably inferior. As Charles Elton has said: "The biological cost of modern transport is high."*

In short, we face not only an unfavorable balance between loss and gain in habitat, but an accelerating disorganization of those unknown controls which stabilize the flora and fauna, and which, in conjunction with stable soil and a normal regimen of water, constitute land-health.

THE HUMAN BACKGROUND. Behind both of these trends in the physical status of the landscape lies an unresolved contest between two opposing philosophies of farm life. I suppose these have to be labeled for handy reference, although I distrust labels:

1. *The farm is a food-factory,* and the criterion of its success is salable products.
2. *The farm is a place to live.* The criterion of success is a

*Journal of Animal Ecology 13, No. 1 (1944), pp. 87–88.

harmonious balance between plants, animals, and people; between the domestic and the wild; between utility and beauty.

Wildlife has no place in the food-factory farm, except as the accidental relic of pioneer days. The trend of the landscape is toward a monotype, in which only the least exacting wildlife species can exist.

On the other hand, wildlife is an integral part of the farm-as-a-place-to-live. While it must be subordinated to economic needs, there is a deliberate effort to keep as rich a flora and fauna as possible, because it is "nice to have around."

It was inevitable and no doubt desirable that the tremendous momentum of industrialization should have spread to farm life. It is clear to me, however, that it has overshot the mark, in the sense that it is generating new insecurities, economic and ecological, in place of those it was meant to abolish. In its extreme form, it is humanly desolate and economically unstable. These extremes will some day die of their own too-much, not because they are bad for wildlife, but because they are bad for farmers.

When that day comes, the farmer will be asking us how to enrich the wildlife of his community. Stranger things have happened. Meanwhile we must do the best we can on the ecological leavings.

The Land-Health Concept and Conservation

Leopold left this extraordinary essay in pencil draft at his death. Written in December 1946, it presents the idea of land health with greater clarity and detail than do any of his other writings. It is published here for the first time. Leopold presents land health not just as a desirable attribute of a landscape but as a much-needed focus for conservation work, and he issues a plea to his fellow ecologists to join him in offering their best guesses about the requisites for land health. The need for action, he says, is urgent, and conservation workers cannot wait until ecologists have all the answers. Leopold also presents here his strongest assertion that land-owners, particularly farmers, ought to shoulder affirmative duties to promote the common good. He ends the essay with a subject that he often addressed in unpublished manuscripts but never really dealt with in print—the need to stabilize "human density" and the possibility that natural forces might keep human numbers in check if social forces do not.

AUGUSTE COMTE, and later Herbert Spencer, pointed out that there is a natural sequence in the development of the sciences, and that this sequence represents a gradient from the simple toward the complex. Spencer's sequence was: physics—chemistry—biology—psychology—sociology.

According to this sequence, ecology, the sociology of the biota, will be the last science to achieve the stage of predictable reactions. This expectation presents a peculiar dilemma, because there is urgent need of predictable ecology at this moment. The reason is that our new physical and chemical tools are so powerful and so widely used that they threaten to disrupt the capacity for self-renewal in the biota. This capacity I will call land-health.

The symptoms of disorganization, or land sickness, are well known. They include abnormal erosion, abnormal intensity of floods, decline of yields in crops and forests, decline of carrying capacity in pastures and ranges, outbreak of some species as pests and the disappearance of others without visible cause, a general tendency toward the shortening of species lists and of food chains, and a world-wide dominance of plant and animal weeds. With hardly a single exception, these phenomena of disorganization are only superficially understood.

George P. Marsh, in *The Earth as Modified by Human Action* (1874), was one of the first to sense that soil, water, plants, and animals are organized collectively in such a way as to present the possibility of disorganization. His case histories describe many degrees of biotic sickness in many geographic regions. They are probably the ultimate source of the biotic ideas now known as conservation.

One might offer an ironic definition of conservation as follows: Conservation is a series of ecological predictions made by beginners because ecologists have failed to offer any.

Need I stop to prove this? The names of Theodore Roosevelt, Gifford Pinchot, William T. Hornaday, Hugh H. Bennett, and Jay N. Darling seem to spring out of recent American history with an emphatic reply. This paper is, in substance, a plea for ecological prediction by ecologists, whether or no the time is ripe. If we wait for our turn in the Spencerian sequence, there will not be enough healthy land left even to define health. We are, in short, land-doctors forced by circumstance to reverse the logical order of our service to society. No matter how imperfect our present ability, it is likely to contribute something to social wisdom which would otherwise be lacking.

Conditions Requisite for Land-Health

I have no illusion that the thousands of ecological questions raised by modern land-use can all be assessed by ecologists. What I mean by "prediction" is a shrewd guess on just one basic question: What are the probable conditions requisite for the perpetuation of the biotic self-renewal or land-health? This would define a goal for conservationists to strive toward. They now have no basic goal bracketing all component groups. Each group has its own goal, and it is common knowledge that these conflict and nullify each other to a large degree.

I will record my own guess first as a figure of speech. The biotic clock may continue ticking if we:

 1. Cease throwing away its parts.

2. Handle it gently.
3. Recognize that its importance transcends economics.
4. Don't let too many people tinker with it.

The Integrity of the Parts

Paleontology teaches us that most land was stable, at least in terms of time scales applicable to human affairs, up to the point at which fauna, flora, soil, or waters were radically modified for human use. Disorganization seldom preceded the wholesale conversion of land with modern tools. It is necessary to suppose, therefore, that a high degree of interdependence exists between the capacity for self-renewal and the integrity of the native communities.

To cite a case: Evolution made few changes in the species list of Europe and America since the last glaciation, nor have the soil or water systems changed materially. Communities were pushed around by climatic cycles, but they did not disappear, and their membership remained intact. The big changes in fauna, flora, soil, and water have all occurred in the last few centuries. We must assume, therefore, that some causal connection exists between the integrity of the native communities and their ability for self-renewal. To assume otherwise is to assume that we understand the biotic mechanisms. The absurdity of such an assumption hardly needs comment, especially to ecologists.

There are, of course, practical limits of both time and space which curtail the degree to which the species list can be returned in settled regions. No one debates the removal of the buffalo or the pigeon from the cornbelt. But we are today extinguishing many species, or relegating them to national parks, on grounds that are ecologically false. Thus the timber wolf, already extinguished over most of the West, is at the point of being extinguished in the Lake States, with official sanction and in fact subsidy, because he eats deer. The assumption is that rifles can trim the deer herd, but the fact is that in Wisconsin and Michigan at least, the deer herd is trimming us. Not only are deer nullifying the reforestation program, but they are tending to eliminate at least three tree species from the future forest: white cedar, hemlock, and yew. The proportion of white pine is being lowered in many localities. The effect of excess deer on lesser vegetation, on other animals, and ultimately on soil, is not known, but it may be large. It has been suggested that the snowshoe hare, under the impact of overbrowsing by deer, ceases to exhibit cyclic population behavior, and that the ruffed grouse is injuriously affected through depletion of its food and cover plants.

Here then is a chain reaction of unknown length threatening the integrity of the fauna and flora over great areas, and arising from a single error in prediction: that human predation by rifle is the biotic equivalent of wolf predation.

This is one of hundreds of land-use errors, made by laymen-administrators in the name of conservation, and all based on the assumption that we are at liberty to prune the species list of members considered "useless," harmful, or unprofitable.

That we must alter the distribution and abundance of species before we understand the consequences of doing so is taken for granted. These modifications are reversible, and hence not very dangerous. But extirpation is never reversible. It is already too late to restore the wolf to the western deer ranges because the indigenous races are extinct.

Closely related to the needless pruning of species lists is the question of their needless enlargement by the importation of exotics. Space forbids my covering this. I will only say that the idea of preference for natives hardly exists in fish management, agronomy, and horticulture, and has only a tenuous hold in game management, forestry, and range management. Soil management is just discovering that there is a soil fauna and a soil flora.

Violence in Land-Use

All land must be converted, either in its plant successions, topography, or water relations, before it can support an industrial economy. My guess here is that the less violent these conversions, the more likely they are to be durable, and the less likely they are to exhibit unforeseen repercussions.

A veritable epidemic of violence prevails at the present moment in the field of water management. Flood-control dams, hydro electric dams, channelization and dyking of rivers, watershed authorities, drainages, lake outlet controls, and impoundments are running riot, all in the name of development and conservation. I am not wise enough to know which of these conversions are ecologically sound, but the most superficial observer can see that:

1. Most of them deal with symptoms, not with organic causes.
2. Their promoters are innocent of (or oblivious to) the principle that violence is risky.
3. Many of them involve irreversible changes in the organization of the biota.
4. Collectively, their use of economic arguments is naive. In one case, economic advantage is held to supersede all opposing considerations; in the next "intangible" benefit is held to supersede all economics.
5. In all of them, control of nature by concrete and steel is held to be inherently superior to natural or biotic controls.
6. In all of them, the economic products of violence are held to be more valuable than natural products.

The philosophy of violence extends far beyond water management. The reckless use of new poisons in agronomy, horticulture, wildlife control, fish management, forestry, and soil fumigation is well known. Poisons for public health are no novelty. Poisons to offset pollution in lakes and rivers are no novelty. Again I am not wise enough to say which of these violent treatments are sound, but it is obvious that the same doubts arise: They deal with symptoms; their promoters are innocent of probable repercussions; they involve many irreversible changes; because they are quicker than biotic controls, they are assumed to be superior to them.

Esthetics

The biota is beautiful collectively and in all its parts, but only a few of its parts are useful in the sense of yielding a profit to the private landowner. Healthy land is the only permanently profitable land, but if the biota must be whole to be healthy, and if most of its parts yield no salable products, then we cannot justify ecological conservation on economic grounds alone. To attempt to do so is sure to yield a lop-sided, and probably unhealthy, biotic organization.

Herein lies the tragedy of modern land-use education. We have spent several generations teaching the farmer that he is not obligated to do anything on or to his land that is not profitable to him as an individual. We can thank his neglect and inertia, and perhaps the hollow sound of our own voice, for the survival of such useless plants and animals, and such natural soils and waters, as remain alive today.

We have rationalized this fallacy by relegating the conservation of the merely beautiful to the state. We can thank this subterfuge for our national parks, forests, and a sprinkling of wilderness areas, but we can also thank it for a million farmers who year-by-year grow richer at the bank, poorer in soil, and bankrupt in spiritual relationships to things of the land.

The divorcement of things practical from things beautiful, and the relegation of either to specialized groups or institutions, has always been lethal to social progress, and now it threatens the land-base on which the social structure rests. The fallacy has its roots in an imperfect view of growth. All sciences, arts, and philosophies are converging lines; what seems separate today is fused tomorrow. Tomorrow we shall find out that no land unnecessarily mutilated is useful (if, indeed, it is still there). The true problem of agriculture, and all other land-use, is to achieve both utility and beauty, and thus permanence. A farmer has the same obligation to help, within reason, to preserve the biotic integrity of his community as he has, within reason, to preserve the culture which rests on it. As a member of the community, he is the ultimate beneficiary of both.

Human Density

The trend of animal ecology shows, with increasing clarity, that all animal behavior-patterns, as well as most environmental and social relationships, are conditioned and controlled by density. It seems improbable that man is any exception to this rule.

It is almost trite to say that the ecological state called civilization became possible at a certain minimum density-threshold. It seems equally probable that above a certain maximum density its benefits begin to cancel out, and its mechanisms become unstable. Improvements in organization may raise that maximum, but they can hardly abolish it.

I have studied animal populations for twenty years, and I have yet to find a species devoid of maximum density controls. In some species the control mechanism inheres within the species, and operates by eviction and resultant vulnerability to predation (quail, muskrat). In others the control is external (deer), and consists of predation, or starvation if that fails. In all species one is impressed by one common character: If one means of reduction fails, another takes over.

It is possible to interpret the impending disorganization of land as taking over the reducing job after we foiled the normal mechanism by industrialization, medicine, and other devices. There is a striking parallelism between the present world-wide strife, and the social status of an overpopulated muskrat marsh just prior to catastrophe.

In any event it is unthinkable that we shall stabilize our land without a corresponding stabilization of our density. It is notorious that many of the "undeveloped" regions are already overpopulated.

Conclusion

These then are my personal guesses as to the conditions requisite for land-health. Some of them step beyond "science" in the narrow sense, because everything really important steps beyond it. I do not claim that my guesses are objective. They are admittedly wishful. Objectivity is possible only in matters too small to be important, or in matters too large to do anything about.

Afterword

Stanley A. Temple

TODAY, WE take for granted the ease with which landowners wanting to aid wildlife can obtain sound technical assistance. For example, a search for information on just one related subject, bird feeding, produces a list of more than 760 titles in print and directions to more than 1,500 Internet sites. Once primarily an act of conservation, bird feeding has become a popular spectator sport, with a multimillion-dollar industry providing food and feeders to an estimated 52 million American households. Other thriving businesses specialize in publications, products, and services to assist wildlife enthusiasts in the management of their lands. I should know; as a rural landowner who practices conservation and restoration on my own property, I find my mailbox bombarded regularly with promotional material.

Moreover, conservation organizations such as the National Wildlife Federation sponsor significant outreach programs to promote "backyard wildlife," and private consultants providing habitat-management information can make a comfortable living catering to the needs of concerned landowners. Government agencies and land grant universities employ a small army of extension specialists to give advice on wildlife conservation, and an array of government-funded incentive programs encourage landowners to follow that advice. Since the late 1980s, one such program, the U.S. Department of Agriculture's Conservation Reserve Program, has prompted landowners to convert 50 million acres of cropland to long-term wildlife cover. Today, no landowner interested in helping wildlife should fail for lack of either advice or assistance.

It is hard to imagine that in the late 1930s and early 1940s, when

Aldo Leopold was writing the essays collected in this volume, conservation-minded landowners had few places to turn for encouragement and guidance. The science of wildlife management was just beginning to take shape under Leopold's intellectual leadership, and no trained professionals worked yet in the field. That is, there were no professionals in the field until 1933, when Leopold began training them in his revolutionary program at the University of Wisconsin, using as a text his just-published *Game Management,* the world's first book on the subject. There was not even much published information on such now-commonplace activities as bird-watching until *A Guide to Bird Watching* appeared in 1943, written by one of Leopold's students, Joseph Hickey. In the conservation world of the day, a niche was open for Leopold to occupy, and he stepped into it with authority.

As a new professor of game management in the College of Agriculture at one of the country's premier land grant universities, Leopold began at once to develop a wildlife outreach program, in addition to meeting classroom and research obligations. Outreach was particularly appropriate work at the University of Wisconsin, with its strong tradition of providing educational opportunities to ordinary citizens. Leopold began monthly broadcasts over the university extension's radio station, aimed largely at farmers and other rural landowners. These radio talks, beginning with "Building a Wisconsin Game Crop" in 1933, covered the same practical topics as the essays collected here in "A Landowner's Conservation Almanac." The following year, Leopold began offering a "Farmer's Short Course" in wildlife conservation for landowners and their families who came to the university to learn about new developments. As best as we can tell from outlines of the month-long course and Leopold's lecture notes, the subjects he wrote about in the *Wisconsin Agriculturist and Farmer* and other popular publications were first presented in this course.

Like many professors, Leopold also included field demonstrations among his outreach activities. Most faculty members in schools of agriculture carried out field trials of new farming practices on the university's network of experimental farms, where touring farmers could inspect the latest agricultural techniques. Leopold took a distinctly different tack. He convinced the university to let him experiment on abandoned crop fields in its newly acquired, 500-acre Arboretum and Wild Life Refuge. There, Leopold tested not new production methods but new ideas for restoring farmland that traditional agriculture had severely degraded. Leopold described the arboretum's innovative goal in his keynote address at its 1934 dedication: "Our idea, in a nutshell, is to reconstruct, primarily for the use of the University, a sample of original Wisconsin—a sample of what Dane County looked like when our ancestors arrived here during the 1840s." The arboretum would, he hoped, "serve as a benchmark, a starting point, in the long and laborious job of building a permanent and mutually beneficial relationship between civilized men and a civilized landscape."

Leopold also used his own rural property in Sauk County as a now-famous "sand-county" proving ground for his ideas on ecological restoration. Other field trials of wildlife conservation practices took place under Leopold's guidance on privately owned farms, some of which are featured in his essays: the Riley Game Cooperative, the Faville Grove Wildlife Experiment Area, and the Coon Valley Erosion Project. Ecological restoration has become legitimate today, even fashionable. But Leopold no doubt raised eyebrows among his colleagues when, as an upstart professor in the College of Agriculture, he demonstrated how to take land *out* of agricultural production.

The publication of popular essays in the *Wisconsin Agriculturist and Farmer,* the *Wisconsin Conservation Bulletin,* and other magazines was the third component of Leopold's ambitious outreach program, an endeavor for which he clearly had high hopes. Those hopes were not always realized, however. An early project to write and publish a "Farm Manual" foundered for lack of support. Leopold's series in the *Wisconsin Agriculturist and Farmer,* after thriving for a time, was terminated by the editor, apparently because it began to swerve too far out of the game-crop furrow into the wildlife field. The Wisconsin Agricultural Society did publish a pamphlet collection of some of Leopold's *WAF* essays, but several potential publishers, including the College of Agriculture and the Wisconsin Conservation Department, turned down the chance to publish the more comprehensive handbook for rural land management that Leopold proposed.

To be fair, Leopold's editors and prospective publishers probably assessed accurately the potential audience for his work at the time. The market for the kind of advice Leopold offered was much smaller in the 1930s and 1940s than it is today. Luckily for Leopold and now for us, he found in the University of Wisconsin a sympathetic patron willing to support his pioneering efforts to educate and enlighten the landowning public, unconstrained by the policies of special interest groups or government agencies.

Were Leopold at work today, his message would find a larger and more eager audience. In fact, many of the efforts he initiated have grown and flourished over the years. The direct descendant of Leopold's radio program still airs on the University of Wisconsin's radio station. The short course in farmstead wildlife conservation that Leopold launched in 1934 is still offered by the department he founded, and its audience and objectives remain essentially the same. Similar courses are now taught on many other campuses around the country. The University of Wisconsin's arboretum has become a center for generating and disseminating information about ecological restoration. Further, Leopold's ideas about restoring wildlife habitats have inspired a new generation of rural landowners who, like Leopold, have dedicated themselves to healing retired farmland. Leopold's "shack" has become a shrine for conservationists and restorationists, and his children and grandchildren carry on his vision through the work of the Aldo Leopold Foundation. And now, with this volume, a comprehensive selection of Leopold's outreach publications is finally available to a broader public.

Given the changes that have occurred since Leopold's day in the science of conservation and in the knowledge, attitudes, and values of many landowners, it is appropriate to ask: How well have Leopold's messages withstood the test of time? Do they remain pertinent to the conservation challenges of today?

Leopold played several roles as author of these articles: interpreter, advisor, critic, and philosopher. As interpreter, he explained the natural history and ecology of Wisconsin for rural landowners who may have known little about the natural processes taking place on their lands and how their activities affected them. As innovative advisor, he suggested simple but effective practices to enhance the capacity of farmland to support wildlife. As critic, he challenged contemporary agricultural practices and the institutions that fostered them. Finally, Leopold the philosopher infused all his outreach writings with his ecocentric view of responsible land use: his now-famous "land ethic."

Leopold the Interpreter

Leopold's skills as a naturalist were such that few errors of fact crept into his *descriptions* of farm wildlife. Indeed, I was unable to find a single instance in the essays in "A Landowner's Conservation Almanac" in which subsequent fieldwork has invalidated his observations. Ecology, though, was still a young science when Leopold wrote, and it is thus not surprising that some of his *interpretations* of natural phenomena have been supplanted by later ecological research.

Leopold was willing to use new ecological theory to guide his recommendations, even theory that was tentative and little tested. His essay "The Land-Health Concept and Conservation" was in its essence, as he put it, "a plea for ecological prediction by ecologists, whether or no the time is ripe." Leopold used caution in presenting his own predictions in that essay, labeling them as mere "personal guesses" about the basis of land health, but he might rightly have been more confident. His guesses for the most part were remarkably correct (sometimes even prescient), and they have withstood the test of time.

In other essays, Leopold also appeared—with the benefit of hindsight—overly cautious about ecological reactions. For example, he stated tentatively that fire "may be beneficial to prairie flowers" in the essay "Wildflower Corners" and failed to mention the role of fire at all in "Roadside Prairies." In "From Little Acorns," he correctly interpreted the relationship between fire suppression and invasion of prairies by woody vegetation, yet he hesitated to state what he must have strongly suspected and what ecological studies have subsequently verified: that periodic fire is crucial to the maintenance of remnant prairies as diverse plant and animal communities.

In a few cases, Leopold does appear to have guessed wrong. For example, in "The Farmer as a Conservationist," he speculated about population cycles in rabbits and grouse: "I suspect that cycles are a disorder of animal populations, in some way spread by awkward land-use." Ecologists now suspect just the opposite: cycles tend to break down with human-caused habitat alterations. Nonetheless, such erroneous guesses were a rarity.

Leopold's appraisals of emerging conservation issues were often prophetic. In "The Outlook for Farm Wildlife," he predicted some of the challenging wildlife problems that conservationists would face in the decades ahead. In 1945, a year before DDT came into widespread use as an agricultural insecticide, Leopold warned: "Modern chemistry is developing controls which may be as dangerous as the pests themselves. (Example: DDT.)." He also forecast accurately the expanding threats to native wildlife that occur when global transport brings exotic species into new areas. Invasive exotic plants and animals are now second only to habitat loss as a cause of declining species. Finally, Leopold anticipated that an increasing number of native species, such as the white-tailed deer, would become overabundant and pestilential because of anthropogenic changes visited on natural landscapes.

Leopold the Advisor

Leopold was a master at making straightforward, practical recommendations for managing farm wildlife, and landowners who implemented his recommendations probably achieved Leopold's central goal: to promote farm game and other wildlife that, with care, could thrive in a fragmented, pastoral landscape. By practical necessity, Leopold ignored the needs of species that require large, unbroken habitats, whether forests or grasslands. Moreover, even in the case of farm wildlife, a reader today should remember that Leopold made his recommendations in the context of rural America in the 1930s and 1940s, a landscape that has markedly changed in the decades since then.

The small family farm, a main target of Leopold's recommendations, is no longer the predominant form of land tenure across the Midwest, and much of the marginal farmland that Leopold addressed is no longer in production. Highly erodible slopes today are rarely plowed and are often planted with permanent cover; livestock is rarely pastured in woods or wetlands; poor soils are seldom cropped; and the rate of wetland drainage has slowed significantly. In many cases, retired farmland is controlled by owners who have no plans to use it to produce income. Instead, they are increasingly interested in following Leopold's lead and restoring the land to a more healthy, natural state.

Leopold would probably be pleased with changes such as these that have taken place on the unhealthy rural landscapes he knew. He would not be pleased, however, with many other changes. On much prime farmland, the intensity of agriculture has increased. Clean farming, which Leopold decried, has become more thorough on many lands still in production. Most active farms, family-owned as well as corporate, now depend heavily on a witch's brew of chemicals to maintain their productivity. Many widely used pesticides are toxic to farm wildlife, and runoff from heavily fertilized fields contributes to the eutrophication of waterways. New varieties of crops have contributed to problems unheard of in Leopold's time. Rapidly maturing varieties of alfalfa, for example, are now cut so early in the year that heavy losses of nesting birds are inevitable. Changes in livestock operations have resulted in fewer pastures, where grassland wildlife often thrives, and more mowed hay fields and row crops, where it does not. Confinement livestock operations and huge feedlots now create waste-handling problems on scales Leopold never knew while depriving fields of the animal manure that once promoted soil fertility.

Despite these changes, Leopold's prescriptions remain sound. The private landowner interested in producing game can still benefit from lessons learned at Riley and at Faville Grove. The practices Leopold instituted on those properties increased game-bird populations, and in most settings they would do so today. Regrettably, we do not know how his prescriptions affected a broader cross section of wildlife species. No one at the time kept good records, and the experiments at Riley and Faville Grove were discontinued after Leopold's time.

Leopold's advice on providing food and cover and tolerating predators is also as good today as it was then. In some cases, though, we have become more sophisticated in our techniques. The modern bluebird house is no longer a section of hollow limb with a floor and a roof attached; it is a cleverly engineered, often mass-produced structure designed to foil house sparrows and European starlings. Happily, the number of nest boxes available today for bluebirds and other cavity nesters would astound Leopold.

The restoration of prairies and oak savannas on retired croplands and degraded woodlands has become a cause célèbre among modern-day restorationists. Leopold would have been amazed at how many of the "wildflower corners" and "roadside prairies" of his time have expanded to fill entire fields. Nowhere would the change have pleased him more than on the lands surrounding his shack. Guided by his philosophies, landowners around Leopold's sandy farm have pooled nearly a thousand acres, in various stages of restoration, to form the Aldo Leopold Memorial Reserve. The arboretum that Leopold inspired has flourished, and most of his goals for the property are still pursued.

Leopold the Critic

Throughout his career, Leopold never hesitated to criticize public policies that harmed wildlife and natural ecosystems, just as he was quick to praise programs that benefited them.

In "Be Your Own Emperor," Leopold took a poke at the notion that government could solve conservation problems merely by passing laws. In "The Land-Health Concept and Conservation," he took issue with the New Deal's reactive approach to treating symptoms of "land sickness" via purchase, prohibition, and damage repair. Instead, Leopold favored programs that got landowners actively involved in preventing damage and promoting land health. Accordingly, he probably would have supported many of today's federal programs. "Swampbuster" laws that encourage farmers to leave wetlands undrained; the Conservation Reserve Program, which shifts marginal cropland to long-term cover; the Wildlife Habitat Incentives Program, which promotes habitat restoration on private lands: all seem compatible with Leopold's view of the government's rightful role in conservation. Today's emphasis on landscape-scale conservation planning also fits well with Leopold's ideas about cooperation among landowners to accomplish goals that would be impractical on a farm-by-farm basis, although as Leopold no doubt would note, the subdivision of failed farms into rural homes, even when occupied by sympathetic landowners, makes such large-scale projects more challenging.

Among the essays gathered here, the most trenchant criticism appears in "What Is a Weed?" In it, Leopold decries "propaganda" that encouraged landowners to develop unwarranted prejudices against native species and to damage their lands. More than any other, this essay reveals Leopold the critic at his sardonic best. Were he alive today, he would very likely also be a vocal critic of the Wise Use Movement, which uses propaganda about private property rights to defend environmentally damaging land-use practices.

The academic community, of which he was then a part, did not escape the barbs of Leopold the wry critic. He chided ecologists for being too reluctant to descend from their ivory towers and apply their new science to conservation. (In the 1970s, similar criticisms gave rise to a new academic discipline, conservation biology, with Leopold as its unofficial patron saint.) Leopold took agricultural colleges and their extension programs to task for promoting "slick and clean" land-use practices that destroyed wildlife habitat on farms. So disturbed was Leopold with this form of propaganda that his bottom line was particularly stark: "The land-philosophy of agricultural schools and extension agencies must be turned inside out." Leopold would probably join those who criticize colleges of agriculture today for promoting chemical solutions to agricultural problems while ignoring or denying the overwhelming evidence of harm.

Leopold the Philosopher

The most enduring features of Leopold's essays on farmland wildlife are the underlying philosophies that motivated him to write so prolifically and eloquently about what he called "the oldest task in human history," living on a piece of land without despoiling it.

Leopold believed firmly in active management, rather than passive protection, as the appropriate approach to conservation on human-dominated landscapes. This philosophy pervaded all of his outreach publications.

The essay "Game Management: A New Field for Science" introduced briefly what his book *Game Management* would document in detail: wildlife populations and wildlife habitats could be "grown" by the same type of active, creative interventions that had given rise to the modern farm landscapes so lacking in wildlife. Extending the agricultural analogy, Leopold often referred to wildlife as a "crop" during the years in which he focused intently on making the land produce harvestable populations of game.

By the time he wrote the essays included in "A Landowner's Conservation Almanac," however, Leopold had largely abandoned the agricultural model, even as he was trying to persuade an agricultural audience to become wildlife managers. His focus was still on active management, but his aims included more than simply producing crops of harvestable game. His objects of care had expanded to include the entire wildlife community, including plants.

When he wrote "The Farmer as a Conservationist," Leopold was intent on converting the rural landowner to a well-rounded conservationist, not just a producer of game. His creek, Leopold wrote, should be "unstraightened," the "creek banks . . . wooded and ungrazed"; hollow trees were to be "left for the owls and squirrels," downed logs, "for the coons and fur-bearers." Martin houses and feeding stations were to be provided, the roadside was to be "a refuge for the prairie flora," and the landowner should be "moved by a positive affection for the fauna and flora as a whole." Wildlife was to be encouraged through "a deliberate effort to keep as rich a flora and fauna as possible, because it is 'nice to have around.'"

In "Biotic Land-Use" and "The Land-Health Concept and Conservation," Leopold made his philosophical plea for all landowners, not just farmers, to become responsible for "land-health." In the latter, his prescription for healthy land is simple and straightforward: maintain the biological diversity of the land; use the land in a sustainable, "less violent" way; recognize the land for its beauty as well as its utility; and keep the human population within the land's carrying capacity. Today, this prescription sounds a lot like the "new" approach conservationists have been fostering under the banner of ecosystem management. Perhaps its time may finally have arrived, more than half a century after Leopold first proposed it.

If so, we can thank Aldo Leopold, ecologist, philosopher, and land doctor, for showing us the way.

Acknowledgments

W E T H A N K the Aldo Leopold Foundation for permission to publish the writings of Aldo Leopold included in this book, and we especially thank Nina Leopold Bradley and Buddy Huffaker for their help, support, and advice. At Island Press, our thanks go to Daniel Sayre, editor-in-chief, for immediately seeing the value of this book and expediting its publication; to Jonathan Cobb, executive editor of the Shearwater Books division, for his editorial advice and his management of the book through the production process; to Lea Kleinschmidt, editorial assistant; to William LaDue, production director, and Christine McGowan, production editorial supervisor; to Abigail Rorer, illustrator; and to Pat Harris, copy editor.

The biographical information about Aldo Leopold included in the introduction is based on Curt Meine, *Aldo Leopold: His Life and Work* (Madison: University of Wisconsin Press, 1988). Curt Meine also provided valuable comments on drafts of the introduction and the afterword. Bernard Schermetzler, archival custodian of the Aldo Leopold papers at the University of Wisconsin–Madison, was gracious in helping us locate the Leopold essays, often in multiple drafts, included in this book. Finally, we thank Julie King and Maria Berger of the University of Illinois College of Law for proofreading the transcriptions of the Leopold essays included here.

J. Baird Callicott
Eric T. Freyfogle

Editors' Notes

3-4   *"How do we assess Leopold's words?"* Curt Meine, "The Farmer as Conservationist: Leopold on Agriculture," in *Aldo Leopold: The Man and His Legacy,* ed. Thomas Tanner (Ankeny, Iowa: Soil Conservation Society of America, 1987), p. 51.

4   *"Who is the land?"* Aldo Leopold, "The Role of Wildlife in a Liberal Education," in *The River of the Mother of God and Other Essays by Aldo Leopold,* eds. Susan L. Flader and J. Baird Callicott (Madison: University of Wisconsin Press, 1991), p. 227.

6   *"My father was a Midwesterner..."* Donald Dale Jackson, "Sage for All Seasons," *Smithsonian,* vol. 29, no. 6 (1998), p. 124.

7   *"the more important and complex task..."* Aldo Leopold, "Wilderness," in *The River of the Mother of God and Other Essays by Aldo Leopold,* eds. Susan L. Flader and J. Baird Callicott (Madison: University of Wisconsin Press, 1987), p. 282.

9   *"I liked corn..."* Aldo Leopold, "Foreword [to *A Sand County Almanac*]" in *Companion to A Sand County Almanac: Interpretive and Critical Essays,* ed. J. Baird Callicott (Madison: University of Wisconsin Press, 1987), p. 282.

10-11   *"I should like to call your attention..."* Alice Harper to E. R. McIntyre, December 15, 1941, Aldo Leopold Papers, University of Wisconsin–Madison.

12   *"I wish you would tell me frankly..."* Aldo Leopold to E. R. McIntyre, June 30, 1942, Aldo Leopold Papers, University of Wisconsin–Madison.

12-13   *"I am sorry..."* E. R. McIntyre to Aldo Leopold, June 30, 1942, Aldo Leopold Papers, University of Wisconsin–Madison.

13   *"I entirely understand..."* Aldo Leopold to E. R. McIntyre, October 13, 1942, Aldo Leopold Papers, University of Wisconsin–Madison.

13   *"there are 20 sketches . . ."* Aldo Leopold to Andrew Hopkins, October 12, 1942, Aldo Leopold Papers, University of Wisconsin–Madison.

14-15   *"a fundamental misconception . . ."* Gifford Pinchot, *Breaking New Ground* (New York: Harcourt, Brace and Co., 1947), p. xix.

15   *"Forestry is not aesthetics . . ."* John D. Guthrie to Aldo Leopold, July 8, 1928, Aldo Leopold Papers, University of Wisconsin–Madison.

21   *"To keep every cog and wheel . . ."* Aldo Leopold, "Conservation," in *Round River,* ed. Luna B. Leopold (New York: Oxford University Press, 1953), p. 147.

24   *"The farmer should know . . ."* Aldo Leopold, "Conservation: In Whole or In Part?" in *The River of the Mother of God and Other Essays by Aldo Leopold,* eds. Susan L. Flader and J. Baird Callicott (Madison: University of Wisconsin Press, 1991), p. 318.

47   *"You certainly have packed . . ."* H. H. Bennett to Aldo Leopold, May 22, 1935, Aldo Leopold Papers, University of Wisconsin–Madison.

229   *"Our idea, in a nutshell . . ."* and *"serve as a benchmark . . ."* Aldo Leopold, "What Is the University of Wisconsin Arboretum, Wild Life Refuge, and Forest Experiment Preserve?" in *The Arboretum at 50* (Madison: University of Wisconsin Arboretum and Wildlife Refuge, 1987), pp. 3, 5.

236   *"the oldest task in human history . . ."* Aldo Leopold, "Engineering and Conservation" in *The River of the Mother of God and Other Essays by Aldo Leopold,* eds. Susan L. Flader and J. Baird Callicott (Madison: University of Wisconsin Press, 1991), p. 254.

About the Contributors

J. BAIRD CALLICOTT is professor of philosophy and religion studies at the University of North Texas. He is author or editor of many books and articles about or by Aldo Leopold, including *The River of the Mother of God and Other Essays by Aldo Leopold; Companion to A Sand County Almanac: Interpretive and Critical Essays; In Defense of the Land Ethic: Essays in Environmental Philosophy;* and *Beyond the Land Ethic: More Essays in Environmental Philosophy.*

ERIC T. FREYFOGLE is the author of *Bounded People, Boundless Lands: Envisioning a New Land Ethic* and *Justice and the Earth.* A native of central Illinois and an active local conservationist, he teaches property, environmental, and natural resources law at the University of Illinois College of Law, where he is the Max L. Rowe Professor.

SCOTT RUSSELL SANDERS is the author of more than dozen books about the human place in nature and community, including *Hunting for Hope; Secrets of the Universe;* and *Staying Put.* He is Distinguished Professor of English and director of the Wells Scholars Program at Indiana University.

STANLEY A. TEMPLE is the Beers-Bascom Professor in Conservation in the department of wildlife ecology at the University of Wisconsin-Madison. His position is the direct descendant of the one originally held by Aldo Leopold, and in this capacity Temple now teaches the courses first developed and taught by Leopold. Among his many contributions to the conservation field, Temple has been president of the Society for Conservation Biology and chairman of the Wisconsin chapter of The Nature Conservancy.